ABDUCTIVE REASONING

SYNTHESE LIBRARY

STUDIES IN EPISTEMOLOGY,

LOGIC, METHODOLOGY, AND PHILOSOPHY OF SCIENCE

VOLUME 330

ABDUCTIVE REASONING

LOGICAL INVESTIGATIONS INTO DISCOVERY AND EXPLANATION

by

ATOCHA ALISEDA
National Autonomous University of Mexico

A C.I.P. Catalogue record for this book is available from the Library of Congress.

ISBN-10 1-4020-3906-9 (HB)
ISBN-13 978-1-4020-3906-5 (HB)
ISBN-10 1-4020-3907-7 (e-book)
ISBN-13 978-1-4020-3907-2 (e-book)

Published by Springer,
P.O. Box 17, 3300 AA Dordrecht, The Netherlands.

www.springer.com

Printed on acid-free paper

Printed in the Netherlands.

To Rodolfo

Contents

Foreword

Many types of scientific reasoning have long been identified and recognised as supplying important methodologies for discovery and explanation in science, but many questions regarding their logical and computational properties still remain controversial. These styles of reasoning include induction, abduction, model-based reasoning, explanation and confirmation, all of them intimately related to problems of belief revision, theory development and knowledge assimilation. All of these have been addressed both in the philosophy of science and in the fields of artificial intelligence and cognitive science, but their respective approaches have been in general too far apart to leave room for an integrated account.

My general concern in this book is scientific discovery and explanation. The point of departure is to address an old but still unsettled question: does scientific methodology have a logic? There have been several conflicting stances about how to pose this question in the first place - not to mention the solutions - each of which rests on its own assumptions about the scope and limitations of scientific methodology and also in their attitude to logic. This question has been posed within several research traditions and for a variety of motivations and purposes, and that fact has naturally shaped the way it has been tackled. In this respect the answer is often already implicit in the very posing of the question itself, which in any case is not a yes-no matter.

Thus a closer look at aspects of this question is in order. While it is clear to everyone that scientific practice indeed involves both processes of discovery and explanation, the first point of disagreement is concerned with the proper scope of the methodology of science. In the predominant view of XXth century philosophy of science, creativity and discovery is simply out of bounds for philosophical reflection, and thus the above question is focused mainly on issues of explanation and testing. It is well known that great philosophers and mathematicians have been brilliant exceptions in the study of discovery in science, but their contributions have set no new paradigms in the methodology of science, and instead have inspired research in cognitive science and artificial intelligence. There is indeed a rapidly growing research on issues of computational scientific discovery, which is full of computer programs which challenge the philosophical claim that the so called 'context of discovery' does not allow for any formal treatment. Thus, in this tradition discovery is part of scientific methodology, on a par with explanation.

As for the place of logic, there are also a variety of instances of what 'logic' amounts to in posing the above question. On the one hand, in XXth century positivist philosophy of science deductive logic was the predominant formal framework to address issues of explanation and evaluation. From the start it was clear that inductive logic had too little on offer for a proper logical analysis of scientific methodology, and this provided the leeway for alternative proposals, such as Popper's conjectures and refutations logical method. Computer oriented research, on the other hand, identifies logic with 'pattern seeking methods', a notion which fits very well their algorithmic and empirical approach to the above question. In any case, the use of the term 'logic' regarding discovery has had little to do with providing logical foundations for their programs, either as conceived in the mathematical logical tradition or as in artificial intelligence logical research, both of which regard inference as the underlying logical notion. It is clear that classical logic cannot account for any kind of ampliative reasoning, and so far as the logical tools developed in artificial intelligence are concerned, not many of them have been applied to issues of discovery in the philosophy of science.

Aim and Purpose

In this book I offer a logical analysis of a particular type of scientific reasoning, namely abduction, that is, reasoning from an observation to its possible explanations. This approach naturally leads to connections with theories of explanation and empirical progress in the philosophy of science, to computationally oriented theories of belief change in artificial intelligence, and to the philosophical position known as Pragmatism, proposed by Charles Peirce, to whom the term abduction owes its name. The last part of the book is concerned with all these applications.

My analysis rests on several general assumptions. First of all, it assumes that there is no single logical method in scientific practice in general, and with respect to abduction in particular. In this my view is pluralistic. To be sure, abduction is not a new form of inference. It is rather a topic-dependent practice of explanatory reasoning, which can be supported by various notions of inference, classical and otherwise. By this assumption, however, I do not claim it is possible to provide a logical analysis for all and every part of scientific inquiry. In this respect, my enterprise is modest and has no pretensions that it can offer either a logical analysis of great scientific discoveries, or put forward a set of logical systems that would provide general norms to make new discoveries. My aim is rather to lay down logical foundations in order to explore some of the formal properties under which new ideas may be generated and evaluated. The compensation we gain from this very modest approach is that we can gain some insight into the logical features of some parts of the scientific discovery and explanation processes. This is in line with a well-known view in the philosophy of science, namely that phenomena take place within traditions, something which echoes Kuhn's distinction between normal and revolutionary science. Hence, another general assumption is that a logical analysis of scientific discovery of the type I propose is for normal science, not denying there may be a place for some other kind of logical analysis of revolutionary science, but clearly leaving it out of the scope of this enterprise.

Another general assumption is that the methodological distinction between the contexts of discovery and justification is an artificial one. It can be dissolved if we address abduction as a process rather than as a ready-made product by itself for us to study. Historical as well as computationally oriented research of scientific discoveries show very clearly that new ideas do not just come out of the blue (even though there may be cases of sudden flashes of insight), and that the process of discovery often involves a lot of explanation, evaluation and testing on the way, too. So, it may be possible to address the justification part of science all by itself, as contemporary philosophy has done all along, but once discovery issues come into play, an integrated account of both is needed.

Content Description

This book is divided into three parts: (I) Conceptual Framework, (II) Logical Foundations, and (III) Applications, each of which is briefly described in what follows.

In part I, the setting for the logical approach taken in this book is presented. One the one hand, chapter 1 offers a general overview of the logics of discovery enterprise as well as the role of logic in scientific methodology, both in philosophy of science and in the fields of artificial intelligence and cognitive science. The main argument is that logic should have a place in the normative study in the methodology of science, on a par with historical and other formal computational approaches. Chapter 2 provides an overview of research on abduction, showing that while there are general features and in most cases the main inspiration comes from the American pragmatist, Charles S. Peirce, each approach has taken a different route. To delineate our subject more precisely, and create some order, a general taxonomy for abductive reasoning is then proposed. Several forms of abduction are obtained by instantiating three parameters: the kind of reasoning involved (e.g., deductive, statistical), the kind of observation triggering the abduction (novelty, or anomaly with respect to some background theory), and the kind of explanations produced (facts, rules, or theories).

In part II, the logical foundations of this enterprise are laid down. In chapter 3, abduction is investigated as a notion of logical inference. It is shown that this type of reasoning can be analyzed within various kinds of logical consequence as the underlying inference, namely as classical inference (backwards deduction), statistical or as some type of non-monotonic inference. The logical properties of these various 'abductive explanatory kinds' are then investigated within the 'logical structural analysis', as proposed for non-monotonic consequence relations in artificial intelligence and dynamic styles of inference in formal semantics. As a result we can classify forms of abduction by different structural rules. A computational logic analysis of processes producing abductive inferences is then presented in chapter 4, using and extending the mathematical framework of semantic tableaux. I show how to implement various search strategies to generate various forms of abductive explanations. Our eventual conclusion for this part is that abductive processes should be our primary concern, with abductive explanatory inferences as their secondary 'products'.

Part III is a confrontation of the previous analysis and foundations with existing themes in the philosophy of science and artificial intelligence. In particular, in chapter 5, I analyze the well-known Hempelian models for scientific explanation (the deductive-nomological one, and the inductive-statistical one) as forms of abductive explanatory arguments, the ultimate products of abductive reasoning. This then provides them with a structural logical analysis in the style of chapter 3. In chapter 6, I address the question of the dynamics of empirical progress, both in theory evaluation and in theory improvement. I meet the challenge made by Theo Kuipers [Kui99], namely to operationalize the task of 'instrumentalist abduction', that is, theory revision aiming at empirical progress. I offer a reformulation of Kuipers' account of empirical progress into the framework of (extended) semantic tableaux, in the style of chapter 4, and show that this is indeed an appealing method to account for empirical progress of some specific kind of empirical progress, that of lacunae.

The remaining two chapters have a common argument, namely that abduction may be viewed as a process of epistemic change for belief revision, an idea which connects naturally to the notion of abduction in the work of Charles Peirce, and that of belief revision in the work of Peter Gärdenfors, thus suggesting a direct link between philosophy and artificial intelligence. In chapter 7, I explore the connection between abduction and pragmatism, as proposed by Peirce, showing that the former is conceived as an epistemic procedure for logical inquiry, and that it is indeed the basis for the latter, conceived as a method of philosophical reflection with the ultimate goal of generating 'clear ideas'. Moreover, I argue that abduction viewed in this way can model dynamics of belief revision in artificial intelligence. For this purpose, an extended version of the semantic tableaux of chapter 4 provides a new representation of the operations of expansion, and contraction, all of which shapes the content of chapter 8.

Acknowledgements

Abductive Reasoning presents the synthesis of many pieces that were published in various journals and books. Material of the following earlier publications, recent publications or publications shortly to appear has been used, with the kind permission of the publishers:

'Sobre la Lógica del Descubrimiento Científico de Karl Popper', *Signos Filosóficos*, supplement to number 11, vol. VI, January–June, pp. 115–130. Universidad Autónoma Metropolitana. México. 2004. (Chapter 1).

'Logics in Scientific Discovery', *Foundations of Science*, Volume 9, Issue 3, pp. 339–363. Kluwer Academic Publishers. 2004. (Chapter 1).

Seeking Explanations: Abduction in Logic, Philosophy of Science and Artificial Intelligence. PhD Dissertation, Philosophy Department, Stanford University. Published by the Institute for Logic, Language and Computation (ILLC), University of Amsterdam, 1997. (ILLC Dissertation Series 1997–4). (Chapters 2, 3, 4, 5 and 8).

'Mathematical Reasoning Vs. Abductive Reasoning: An Structural Approach', *Synthese* 134: 25–44. Kluwer Academic Press. 2003. (Chapter 3).

'Computing Abduction in Semantic Tableaux', *Computación y Sistemas: Revista Iberoamericana de Computación*, volume II, number 1, pp. 5–13. Centro de Investigación en Computación (CIC), Instituto Politécnico Nacional, México. 1998. (Chapter 4).

'Abduction in First Order Semantic Tableaux', together with A. Nepomuceno. Unpublished Manuscript. 2004. (Chapter 4).

'Lacunae, Empirical Progress and Semantic Tableaux', to appear in R. Festa, A. Aliseda and J. Peijnenburg (eds). *Confirmation, Empirical Progress, and Truth Approximation (Poznan Studies in the Philosophy of the Sciences and the Humanities*, vol. 83), pp. 141–161. Amsterdam/Atlanta, GA: Rodopi. 2005. (Chapter 6).

'Abduction as Epistemic Change: A Peircean Model in Artificial Intelligence'. In P. Flach and A. Kakas (eds). *Abductive and Inductive Reasoning: Essays on their Relation and Integration*, pp. 45–58. Kluwer Academic Publishers, Applied Logic Series. 2000. (Chapters 7 and 8).

'Abducción y Pragmati(ci)smo en C.S. Peirce'. In Cabanchik, S., etal. (eds.). *El Giro Pragmático en la Filosofía Contemporánea*. Gedisa, Argentina. 2003. (Chapter 7).

The opportunity to conceive this book was provided by a series of research stays as a postdoc at the Philosophy Department in Groningen University, during the spring terms of three consecutive years (2000–2002). I thank my home institution, for allowing me to accept the kind invitation by Theo Kuipers. In Groningen I joined his research group *Promotion Club Cognitive Patterns* (PCCP) and had the privilege to discuss my work with its members and guests, notably with David Atkinson, Alexander van den Bosch, Erik Krabbe, Jeanne Peijenburg, Menno Rol, Jan Willem Romeyn, Rineke Verbrugge and John Woods. I am especially grateful to Theo for making me realize of the fertility of logical research on abduction, particularly when it is guided by questions raised in the philosophy of science.

In the Instituto de Investigaciones Filosóficas at the Universidad Nacional Autónoma de México, the *Seminario de Razonadores* ('Reasoners Seminar') has been a wonderful forum over the past six years for the discussion of reasoning themes amongst philosophers, logicians, mathematicians and computer scientists. I wish to mention the following member colleagues, all of which have contributed to my thinking about abduction: José Alfredo Amor, Axel Barceló, Maite Ezcurdia, Claudia Lorena García, Francisco Hernández, Raymundo Morado, Ana Bertha Nova, Samir Okasha, Silvio Pinto, Marina Rakova and Salma Saab.

I also wish to thank the following people for inspiring discussions and generous encouragement towards my abductive research: Hans van Ditmarsch, Peter Flach, Donald Gillies, Michael Hoffmann, Lorenzo Magnani, Joke Meheus, Ulianov Montaño, Angel Nepomuceno, Jaime Nubiola, León Olivé, Sami Paavola, David Pearce, Ana Rosa Pérez Ransanz, Víctor Rodríguez, Víctor Sánchez Valencia†, Matti Sintonen, Ambrosio Velasco, Rodolfo Vergara and Tom Wasow. In particular, I am grateful to Johan van Benthem, for his guidance during my doctoral research at Stanford. Finally, I thank the anonymous referee for the many insightful suggestions to improve this book. Of course, the responsibility of any shortcomings is mine.

Atocha Aliseda,
July 2004,
Tepoztlán, México.

PART I

CONCEPTUAL FRAMEWORK

Chapter 1

LOGICS OF GENERATION AND EVALUATION

1. Introduction

The general purpose of this chapter is to provide a critical analysis on the controversial enterprise of 'logics of discovery'. It is naturally divided into six parts. After this introduction, in the second part (section 2) we briefly review the original heuristic methods, namely *analysis* and *synthesis*, as conceived in ancient Greece. In the third part (section 3) we tackle the general question of whether there is a logic of discovery. We start by analyzing the twofold division between the contexts of discovery and justification, showing that it may be not only further divided, but also its boundaries may not be so sharply distinguished. We then provide a background history (from antiquity to the XIXth century), divided into three periods in time, each of which is characterized by an epistemological stance (infallibilism or fallibilism) and by the types of logic worked out (generational, justificatory inductive logics, non-generational and self-corrective logics). Finally, we motivate the division of this general question into other three questions, namely one of *purpose*, one of *pursuit* and one of *achievement*, for in general, there is a clear gap between the search and the findings in the question of a logic of discovery. In the fourth part (section 4), we confront two foremost views on the logic of discovery, namely those of Karl Popper and Herbert Simon, and show that despite appearances, their approaches are close together in several respects. They both hold a fallibilist stance in regard to the well-foundedness of knowledge and view science as a dynamic activity of problem solving in which the growth of knowledge is the main aspect to characterize. We claim that both accounts fall under the study of discovery –when a broad view is endorsed– and the convergence of these two approaches is found in that neither Simon's view really accounts for the epistemics of creativity at large, nor Popper neglects its study entirely. In the fifth part (section 5), we advance the claim that logic should have a place in the methodology of science, on a pair with historical and computational stances, something that naturally gives place to logical approaches to the logic of discovery, to be cherish in a normative account of the methodology of science. However, we claim that the label 'logics of discovery' should be replaced by 'logics of generation and evaluation', for on the one hand 'discovery' turns out to be a misleading term for the processes of generation of new knowledge and on the other hand, a logic of generation can only be conceived together with an account of processes for evaluation and justification. In the final part of this chapter (section 6), we sum up our previous discussion and advance our general conclusions.

This chapter shows that interest on the topic of the logics of discovery goes back to antiquity and spans over our present days, as it pervades several disciplines, namely Philosophy of Science, Cognitive Science and Artificial Intelligence. The search of a logic of discovery is not a trivial question, in fact is not even a single question, and the words 'logic' and 'discovery' may be interpreted in several many ways. This chapter will serve to set the ground and philosophical motivation behind our main purpose in this book, namely the logical study of abductive reasoning, which we motivate in the next chapter (chapter 2).

2. Heuristics: A Legacy of the Greeks

All roads in the study of discovery lead back to the Greek mathematicians and philosophers in antiquity. In their study of the processes for solving problems, two were the main heuristic strategies, namely *analysis* and *synthesis*. An extensive description of these methods is found in the writings of Pappus of Alexandria (300 A.D.), who follows a tradition found in 'The Treasury of Analysis' (*analyomenos*), a collection of books by earlier mathematicians. The central part of his description reads as follows (Pappi Alexandrini *Collectionis Quae Supersunt*. The translation is from [HR74, Chapter II]):

> *'Now analysis is the way from was is sought – as if it were admitted – through its concomitants (τα ακολουθα, the usual translation reads: consequences) in order to something admitted in synthesis. For in analysis we suppose that which is sought to be already done, and we inquire from what it results, and again what is the antecedent of the latter, until we on our backward way light upon something already known and being first in order. And we call such a method analysis, as being a solution backwards. In synthesis, on the other hand, we suppose that which was reached last in the analysis to be already done, and arranging in their natural order as consequents the former antecedents and linking them one with another, we in the end arrive at the construction of the thing sought. And this we call synthesis'.*

> *'Now analysis is of two kinds. One seeks the truth, being call theoretical. The other serves to carry out what was desired to do, and this is called problematical. In the theoretical kind we suppose the thing sought as being and as being true, and then we*

> *pass through its concomitants (consequences) in order, as though they were true and existent by hypothesis, to something admitted; then, if that which is admitted to be true, the thing sought is true, too, and the proof will be the reverse of analysis. But if we come upon something false to admit, the thing sought will be false, too. In the problematical kind we suppose the desired thing to be known, and then we pass through its concomitants (consequences) in order, as though they were true, up to something admitted. If the thing admitted is possible or can be done, that is, if it is what the mathematicians call given, the desired thing will also be possible. The proof will again be the reverse of analysis. But if we come upon something impossible to admit, the problem will also be impossible'.*

Analysis and synthesis may be thought of as methods running in reverse of each another, provided that all steps in analysis are reversed when used for synthesis. However, this depends on the interpretation given to '$\tau\alpha$ $\alpha\kappa o\lambda ov\theta\alpha$'. While some scholars regard it as 'consequences', suggesting these methods consist entirely of reversible steps, some others (notably [HR76]), interpret it as 'concomitants', meaning *'almost any sort of going together. Hence Pappus' general description of analysis depicts it consistently as a search for premisses, not as drawing of consequences.'* [HR76, p. 255].

Moreover, the methods of analysis and synthesis do not exist in isolation, but only make sense in combination. It is not enough to arrive at something known to be true in order to assert the truth of the initial statement which analysis takes only as presupposed; the method of synthesis has to provide a proof for it. Either the proof is constructed by inverting each step of the analysis (under the 'consequence' interpretation) or the process of analysis may provide material to construct a proof, but is no final warrantee for a successful synthesis.

Both types of analysis begin with a hypothetical statement, but while in the theoretical kind the statement is supposed to be true, in the problematical kind the statement is supposed to be known. Then, a decisive point is reached when something is found to be either admittedly true (or possible) or admittedly false (or impossible). In the first case, the initial statement can find its proof of being true by way synthesis, if (again) the backward and forward steps of these methods are indeed reversible. In the second case however, if by way of analysis we arrive at a statement that is admittedly false (or impossible), this is enough to assert the falsity (or impossibility) of the initial statement. It comes at no surprise that this situation *'is just a special case of the so–called* **reductio ad absurdum***; because in such a case our starting theorem is, beyond doubt,* **false***'*.[Sza74, p. 126].

It is however a matter of debate to what exactly these methods amounted to in antiquity, for the text of Pappus *'does not suffice to reconcile his general description of analysis with his own mathematical practice, or with the practice of other ancient mathematicians'* [HR76, pp. 255–256]. An attempt to provide a modern and formalized account of these heuristic methods has been given by Hintikka and Remes ([HR74, HR76]), who put forward the thesis that analysis is a special case of natural deduction methods. In particular, they propose Beth's method of semantic tableaux as a method of analysis[1]. As it is well–known, semantic tableaux are primarily a refutation method (cf. chapter 4); when an entailment does not hold, open branches of the tableau are records of counter–examples. Otherwise, the tableau itself provides information to construct a sequent style proof. Therefore, the method of analysis provides a definite proof (by reductio) of the falsity of the 'thing sought' in the case of refutation (when counter–examples have been found). As we shall see (cf. chapter 4), our approach to abduction also takes semantic tableaux as its mathematical framework for a logical reconstruction of this type of reasoning. Therefore, the work by Hintikka and Remes may be regarded as a precursor of our own.

3. Is There a Logic of Discovery? Contexts of Research

The literature on the subject of scientific discovery is staggeringly confused by the ambiguity and complexity of the term *discovery*. A discovery of an idea leading to a new theory made in science involves a complicated process that goes from the initial conception of an idea throughout its justification and final settlement as a new theory. These two aspects are just the two extremes in a series of intermediate processes including the entertainment of a new idea, its initial evaluation, which may lead to finer ideas in need of evaluation or may even be replaced by other ideas, calling for modification of the original one. Therefore, we must at least acknowledge that scientific discovery is a process subject to division. This fact naturally confronts us with the problem of how to provide a proper division.

An attempt to supply a division is given by the contemporary distinction between the contexts of discovery and the context of justification, originally proposed by Reichenbach in the 30's [Rei38]. This division often presupposes the latter as dealing exclusively with the 'finished research report' of a theory, and thus certainly leaves ample room to the former. In order to bring some order and clarity into the study of the process of scientific discovery at large, several authors identify an intermediate step between the two extremes, the conception and justification of a new idea. While Savary puts forward the phase of 'working with ideas' [Sav95], Laudan introduces the 'context of pursuit' as a 'nether region' between the two contexts [Lau80, p. 174]. Another dimension in the study of the context of discovery is to distinguish between a narrow and a broad view. While the former view regards issues of discovery as those dealing

[1] Although Beth himself was inspired by the methods of analysis and synthesis to device his method of semantic tableaux, he *'neither connected this logical idea in so many words with the ancient geometrical analysis nor applied his approach to the elucidation of historical problems.'* [HR76, p. 254].

exclusively with the initial conception of an idea, the latter view is that which deals with the overall process going from the conception of a new idea to its settlement as an idea subject for ultimate justification (a distinction introduced by Laudan in [Lau80]).

It is however also a matter of choice to extend the boundaries of the context of justification in order to deal with evaluation questions as well, especially when the truth of a theory is not the sole interest to be searched for. A consequence of this move is the proposal to rename the 'context of justification' as the 'context of evaluation' [Kui00, p.132] or as the 'context of appraisal' [Mus89, p. 20]. Under the latter view, the context of discovery has in turn been relabeled the 'context of invention', in order to avoid the apparent contradiction that arises when we speak of the discovery of hypotheses, as discovery is a 'success word' which presupposes that what is discovered, must be true.

Therefore, the original distinction between the contexts of discovery and justification may not only be further divided, but also its boundaries may not be so sharply distinguished. A separate issue concerned with all these contexts of research is to inquire whether the 'context of discovery' or any other context for that matter is subject to philosophical reflection and allows for logical analysis.

Background History

In what follows, I will offer an overview of the evolution of the enterprise of the logic of discovery (based on [Lau80], [Nic80] and [Mus89]), from antiquity to the XIXth century. This overview will serve us to identify those relevant aspects that have played a role in the study of discovery in order to place them in the research agenda nowadays.

Since antiquity the enterprise of a logic of discovery was conceived as providing an instrument for the generation of genuine new concepts and theories in science. This enterprise was at the core of Aristotle's *Posterior Analytics* as an attempt to provide an account of the discovery of causes by finding missing terms in incomplete explanations[2]. According to some interpretations ([Lau80], [Ke89]), the search for a logic of discovery at this period aimed to capture the 'Eureka moment', and thus discovery was taken in its narrow interpretation.

Thereafter, the project flourished through concrete proposals for finding good and effective methods for making causal discoveries, as witnessed in the work of Bacon, Descartes, Herschel, Boyle, Leibniz and Mill, to name the most representative ones. The division between contexts in scientific research into

[2]On the one hand, Aristotle introduced the notion of *enthymeme* to characterize those incomplete arguments which lack a premise. Once that premise is found, this type of argument is converted into a deductive one. On the other hand, Aristotle characterized two other types of arguments, namely 'apagoge' and 'epagoge', the former later identified with abduction (by Peirce) and the latter generally interpreted as induction.

discovery and justification was not present at this period and the goal was to offer a logic of discovery also providing a solution to the problem of theory justification. The search for such a logic was guided by a strong philosophical ideal, that of finding a universal system, which would capture the way humans reason in science. For this purpose Leibniz proposed a 'characteristica universalis' consisting of a mathematical language, precise and without ambiguities (as opposed to natural languages) in which all ideas could be translated to and by which intellectual arguments were conclusively settled. A consequence of this ideal was the presumption to decide on the validity of chain of arguments via a calculus, the 'calculus ratiocinator'. This naive optimism in what a logic for science could provide, went to extreme of pretending to have access to all scientific truths via this calculi[3].

Leibniz's calculus was a response to Aristotle's deductive syllogistics, which did not provided what they were looking for, a logic that would genuinely generate new sciences. To this end, Bacon proposed the method of 'eliminative induction', a kind of disjunctive syllogism, by which hypotheses are eliminated in favour of the 'true' one. As for Descartes, *'he gave us the handy hints for problem solvers of the Regulae, and a few secretive references to the method of analysis and synthesis proposed by the Greek geometers'* [Mus89, p.18] synthesis analysis. As it turns out, however, their methods did not provided rules to generate genuine new ideas, for they are rather directed to working with already made hypotheses. Bacon's method is indeed a method of selection and Descartes method focuses on ways of decomposing a hypothesis into parts known to be true (analysis into its parts), to later run a synthesis process in order to prove it.

According to our previous discussion of contexts of research, where to locate these methods turns out to be a matter of debate. They clearly belong to the context of justification according to Reichenbach's distinction and under a narrow view of discovery (as pointed out in [Mus89, p. 18]), but to the context of discovery under a broad one. And taking a finer view on contexts, they belong to the context of pursuit or to the context of evaluation or appraisal when the context of justification is enlarged. In any case, it seems that neither of them belongs to the context of invention, as these methods fail to give an account of how is that the hypotheses entertained were initially conceived.

What we find in all these authors is that while their search for a logic of discovery was guided by Leibniz's ideal of a universal calculi, the results obtained from its pursuit, did not live up to this search. Clearly, there is a gap between the 'search' and the 'findings' in the question of a logic of discovery. Therefore it seems useful to divide the ambitious question of whether there is

[3]Nowadays we know this attempt is indeed impossible. There is no general mechanical procedure able to decide on the validity of logical inferences. Predicate logic is undecidable.

a logic of discovery into three of them, namely its *purpose*, *pursuit*, and its *achievement*. That is, to distinguish between what is searched for and what is obtained as a result of that search, in order to evaluate their coherence and thus give an answer to the question of achievement. But before getting into the proposed three-question division, we will present Laudan's [Lau80] critical historical analysis on the question concerning the search of a logic of discovery, in which he is concerned with discovery in its narrow sense, the view that regards issues of discovery as those dealing exclusively with the initial conception of an idea.

Laudan's Analysis

According to this interpretation, two were the main motivations behind the enterprise of a logic of discovery, namely the epistemological problem of the well-foundedness of knowledge and the heuristic and pragmatic problem of how to 'accelerate scientific advance'. Regarding the first one, the prevalent epistemological stance supporting the legitimization of knowledge was an infallibilistic position, one that required an absolute warrant for the newly generated concepts and theories about the world. The goal was to offer a logic of discovery also providing a solution to the problem of theory justification. There was not a unique epistemological position in regard to the justification of theories, but only one of them was compatible with the enterprise of a logic of discovery, namely the one identified with the generational stance[4]. Therefore, infallibilism in turn leads to a generational position, and these two together provide an absolute warrant of our claims about the world via a truth-preserving logic.

The second motivation regarding the problem of how to 'accelerate scientific advance', was the guiding motor of the ultimate goal for a logic of discovery, that of providing *'rules leading to the discovery of useful facts and theories about nature'* [Lau80, p. 175]. The rules proposed, however, were closely linked to another aspect in the evolution of the ideas surrounding discovery, namely with the object of science under examination. Though several characterizations of what scientific inference amounted to, most methodologists up to mid XVIIIth century focused their efforts on the characterization of the discovery of empirical laws, of universal statements concerning observable entities, and thus the

[4]The two opposing groups were known as the 'consequentialists' and the 'generators'. The former group justifies a theory as true through an inspection of (some of) its consequences. In contrast, the latter group renders a scientific theory as established by showing that it follows logically from observational statements. Since the Aristotelian era, however, a logical flaw was acknowledged in the consequentialist position, since it committed the 'fallacy of affirming the consequent' (to argue from the truth of a consequence to the truth of the theory), something which made this view logically inconclusive: *'if theories were to be demonstrably true, such demonstration could not come from any (non exhaustive) survey of the truth of their consequences'* [Lau80, p. 177]. The only circumstances in which infallibilism and consequentialist can agree are when it is possible to (i) provide an exhaustive examination of the consequences of a certain theory and (ii) enumerate and reject exhaustively all possible theories at hand [Lau80, Note 3, p. 183].

focus was for an inductive logic modelling 'enumerative induction'. Scientific research to be analysed had to do with generalizations directly inferred from observations, with physical laws such as 'all gases expand when heated'. It was until the 1750's that several scientists and philosophers searched for a logic to discover explanatory theories, those involving theoretical entities as well (such as found in [Lap04]). In this case, other type of logics were needed, those which involved a more complex mechanism than that of enumerative induction. Usually, these logics involve a background theory (consisting of laws), initial conditions and a relevant observation. The goal of these logical apparatus was to produce a better and truer theory than the background one. The idea of truth-approximation was behind this conception, and accordingly had been labeled 'self-corrective logics'(we shall identify this logics with abduction, cf. chapter 2). To conclude, up to this point, infallibilism and generational logics were hand in hand, together providing a truth preserving logic of discovery to warrant infallible knowledge.

Later on, around mid XIXth century, the enterprise of a logic of discovery in the terms described above was abandoned and replaced by the search of a logic of justification, a logic of post hoc evaluation. The turning point of this move had to do with a major change in view regarding the legitimization of knowledge, namely by a shift from infallibilism to fallibilism. A consequence of this change in epistemological stance is that the sources of knowledge were no longer supported by certain ground, and thus justification could be separated from proof. Thus, the independence of genesis from the justification of theories sets in, for there was no longer a need to unfold the generation of theories in order to characterize their justification, in other words, the enterprise of a logic of discovery becomes 'redundant' from the epistemic point of view, and it is abandoned. Another consequence is that the consequentialist view for the justification of theories becomes in fashion again, as it is now compatible with an fallibilist account, and thus paves the way to the view represented by what later was known as the 'hypothetico-deductive method', which became the standard in theory justification.

Thus, there are clearly three episodes in the evolution of a logic of discovery from antiquity up to the mid XIXth century, characterized as follows:

- Up to 1750:

  Epistemological stance: infallibilism

  Logic: generational and justificatory inductive logic.

- 1750 - 1820:

  Epistemological stance: infallibilism

  Logic: generational and justificatory inductive logic as well as self-corrective logics.

- 1820 - 1850:

  Epistemological stance: fallibilism

  Logic: Nongenerational justificatory inductive logic as well as self-corrective logics.

Underlying Questions

The purpose of a logic of discovery concerns the ultimate goal researchers engaged in this enterprise wish to achieve in the end, and the response (affirmative or negative) to the question of whether there is such a logic is largely grounded on philosophical ideals. The pursuit of a logic of discovery regards the working activities that researchers in the field are engaged in to attain their goals, and the answer to such a question is given in the form of concrete proposals of logics of discovery. Finally, the achievement of a logic of discovery provides an evaluation between previous aspects, of whether what is actually achieved lives up to its purpose. This three-question division allows to evaluate existing proposals on the logic of discovery as to their coherence, and it also provides a finer grain distinction in order to compare confronting proposals, for they may agree on one question while disagree on the other one.

In the first period identified above (from antiquity to mid XIX century), the search for a logic of discovery was guided by a strong philosophical ideal, that of finding a universal system which would capture the way humans reason in science, including the whole spectrum, from the initial conception of new ideas into their ultimate justification. Following the spirit of Leibniz's "Characteristica Universalis", this ideal was the motor behind the ultimate purpose of finding a logic of discovery. As to the question of pursuit, the central method worked out was the modeling of induction, giving place to proposals such as Bacon's 'eliminative induction', which as already mentioned, it is in fact a method for hypothesis selection. Therefore, regarding the question of achievement of a logic of discovery in this period, the products resulting from its pursuit did captured only to a very small extent what the question of purpose was after, and therefore the project was not coherent in regard to what was searched for and what was found in the end. A similar analysis can be given for the second period. In the third period, the original question of purpose is vanished by the advancement of fallibilism; for it was then clear that a universal calculi in which all ideas could be translated to and by which intellectual arguments were conclusively settled, was an impossible goal to achieve. The question to be searched for instead is concerned with a logic of justification, and accordingly, the question of pursuit focused on developing accounts on this matter. There was complete harmony between the search and the findings, and so the question of achievement renders this period as coherent. The abandonment of a logic for the genesis of theories simply deleted the search of such kind of 'discovery' rules from the agenda and instead focused exclusively in mechanisms for the justification of ready–made scientific theories.

However, the aim to find a logic dealing with the conception of new ideas was not completely discarded, as witnessed later in the work of Charles S. Peirce (cf. chapter 2 and 7) and others, but this line of research has remained incoherent as far as the question of achievement is concerned. As absurd as this position seems to be, this is still the ideal to which the "friends of discovery" hold to, but as we shall see in the following corresponding section, a closer analysis into their approach will help to make sense of their stance.

4. Karl Popper and Herbert Simon

From the philosophy of science point of view, Karl Popper (1902-1994) and Herbert Simon (1916-2001) were two well-established scholars who reacted to logical positivism in their own peculiar and apparently opposing views on the logic of discovery. In this section, we will elucidate some aspects of their proposals in order to place them in the present discussion of scientific discovery. On the one hand, when a finer analysis of contexts of research is done, it seems Popper's logic may be considered as part of the context of discovery, and his account on the growth of knowledge by the method of conjectures and refutations, is in accordance with the "Friends of Discovery" mainstream, of which Simon is one of the pioneers. On the other hand, we will unfold Simon's claim that there is a logic of discovery, in order to show that far from being a naive aim to unravel the mysteries of scientific creativity, it is a solid proposal in the direction of a normative theory of discovery processes, grounded on the view of logic as a pattern seeking activity based on heuristic strategies to meet its ends.

The Logic of Discovery

It is an unfortunate yet an interesting fact that Popper's *Logik der Forschung* first published in German in 1934, was translated into English and published twenty five years later as *The Logic of Scientific Discovery*. An accurate translation would have been: *The Logic of Scientific Research*, as found in translations into other languages, such as Spanish (*La Lógica de la Investigación Científica*).

One reason for its being unfortunate lies in the fact that several renowned scholars (Simon and Laudan amongst others) have accused Popper of denying the very subject matter of what the English title of his book suggests, something along the lines of a logical enterprise into the epistemics of scientific theory discovery. These complaints are firmly grounded within the accepted view that for Popper scientific methodology concerns mainly the testing of theories, and this approach clearly leaves outside of its scope issues having to do with discovery. Thus, for those philosophers of science interested in discovery processes as well as in other methods for scientific inquiry outside the realm of justification, it seems natural to leave Popper out of the picture and take the above complaint on the title to be just a confusion originated from its English translation.

However, on a closer look into Popper's philosophy, an additional confusion stirs up when we find (in a publication shortly after) his position beyond justification issues and concerned with the advancement and discovery in science, as witnessed by the following quote: *'Science should be visualized as progressing from problems to problems - to problems of increasing depth. For a scientific theory - an explanatory theory - is, if anything, an attempt to solve a scientific problem, that is to say, a problem concerned with the discovery of an explanation'* [Pop60a, p. 179]. This view is in accord with Simon's famous slogan that 'scientific reasoning is problem solving' made in research in cognitive psychology and artificial intelligence (to be later introduced), a claim also put forward by Laudan in the philosophy of science. Moreover, it seems Popper was happy with the English title of his book, for being such an obsessive proof reader of his own work, he made no remarks about it[5].

Additionally, it is also appealing that in the literature of knowledge discovery in science, especially from a computational point of view, some of Popper's fundamental ideas are actually implemented within the simulation of discovery and testing processes in science[6].

The common view on Popper's position states that issues of discovery cannot be studied within the boundaries of methodology, for he explicitly denies the existence of a logical account for discovery processes, regarding its study a business of psychology. This position is backed up by the following –much too quoted– passage:

> *"Accordingly I shall distinguish sharply between the processes of conceiving a new idea, and the methods and results of examining it logically. As to the task of the logic of knowledge -in contradistinction to the psychology of knowledge - I shall proceed on the assumption that it consists solely in investigating the methods employed in those systematic tests to which every new idea must be subjected if it is to be seriously entertained. Some might object that it would be more to the purpose to regard it as the*

[5]Popper used every opportunity to clarify his claims and terms put forward in his 'Logik der Forschung', as witnessed in a multitude of footnotes and new appendices to 'The Logic of Discovery' and in many remarks found in later publications. In [Pop92, p. 149] he reports (referring to his 'Postscript: After Twenty Years'): *'In this Postscript I reviewed and developed the main problems and solutions discussed in Logik der Forschung. For example, I stressed that I had rejected all attempts at the justification of theories, and that I had replaced justification by criticism'.*

[6]For example, the requirement characterizing a theory as 'scientific' when it is subject to refutation, is translated into the 'FITness' criterion, and plays a role in the process of theory generation, for a proposed theory only survives if it can be refuted within a finite number of instances [SG73].

> *business of epistemology to produce what has been called a 'rational reconstruction' of the steps that have led the scientist to a discovery -to the finding of some new truth ... But this reconstruction would not describe these processes as they actually happen: it can give only a logical skeleton of the procedure of testing. Still, this is perhaps all that is meant by those who speak of a 'rational reconstruction' of the ways in which we gain knowledge.*
>
> *It so happens that my arguments in this book are quite independent of this problem. However, my view of the matter, for what is worth, is that there is no such thing as a logical method of having new ideas, or a logical reconstruction of this process"* [Pop34, P. 134].

First of all, it should come at no surprise that Popper's position holds a place in the discussion of discovery issues, as the objects of his analysis are precisely genuinely new ideas. Moreover, we should emphasize that Popper draws a clear division between two processes; amidst the conception of a new idea, and the systematic tests to which a new idea should be subjected to, and in the light of this division he advances the claim that not the first but only the second one is amenable to logical examination.

Conjectures and Refutations: The Growth of Scientific Knowledge

For Popper the growth of scientific knowledge was the most important of the traditional problems of epistemology [Pop59, P. 22]. His fallibilist position provided him with the key to reformulate the traditional problem in epistemology, which is focused on the reflection on the sources of our knowledge. Laid down this way, this question is one of origin and begs for an authoritarian answer [Pop60b, p. 52], regardless of its answer being placed in our observations or in some fundamental assertions lying at the core of our knowledge. Popper's answer to this question is that we do not know or even hope to know about the sources of our knowledge, since our assertions are only guesses [Pop60b, p. 53]. Rather, the question and its answer should be the following: *"How can we hope to detect and eliminate error? Is, I believe, By criticizing the theories or guesses of others"* [Pop60b, p. 52]. This is the path to make knowledge grow: *"the advance of knowledge consists, mainly, in the modification of earlier knowledge"* [Pop60b, p. 55].

This concern on the growth of knowledge is intimately related to his earlier mentioned view of science as a problem solving activity, and in this respect he writes: *"Thus, science starts from problems, and not from observations; though observations may give rise to a problem, specially if they are unexpected"* [Pop60a, p. 179]. Moreover, rather than asking the question 'How do we jump from an observation statement to a theory?, the proper question to ask is the following: *"How do we jump to from an observation statement to a good theory?' ... By jumping first to any theory and then testing it, to find whether it is good or not; i.e. by repeatedly applying the critical method, eliminating many bad theories, and investing many new ones"* [Pop63, p. 55].

Thus, the intention in both the title and content of Popper's The Logic of Scientific Discovery was to unfold the epistemics of evaluation and selection of newly discovered ideas in science, more in particular, of scientific theory choice. To this end, Popper proposed a rational method for scientific inquiry, the method of conjectures and refutations, which he refined in later publications [Pop63][7]. The motivation behind this method was to provide a criterion of demarcation between science and pseudo-science. Besides its purpose, the logical method of conjectures and refutations is a norm for progress in science, for producing new and better theories in a reliable way. For Popper the growth of scientific knowledge –and even of pre-scientific knowledge– is based on the learning from our mistakes, which according to him is achieved by the method of trial and error. He did not provided a precise procedure to perform scientific progress, leading to better theories, but rather a set of theory evaluation criteria including a measure for its potential progressiveness (its testability) and the condition of having a greater 'empirical content' than the antecedent theory.

Does Scientific Discovery Have a Logic?

In principle, the pioneering work of Herbert Simon and his team share the ideal on which the whole enterprise of artificial intelligence was initially grounded, namely that of constructing intelligent computers behaving like rational beings, something which resembles the philosophical ideal which guided the search for a logic of discovery in the first period identified above (up to the XIX century). However, it is important to clarify on what terms is this ideal inherited in regard to the question of purpose of a logic of discovery, on the one hand, and put to work with respect to the question of the pursuit of such a logic, on the other.

In his essay *'Does scientific discovery have a logic?* Simon sets himself the challenge to refute Popper's general argument, reconstructed for his purposes as follows: *'If 'There is no such thing as a logical method of having new ideas, then there is no such thing as a logical method of having* **small** *new ideas"* [Sim73a, p. 327]. (My emphasis), and his strategy is precisely to show that an antecedent in the affirmative does not commit to an assessment of the consequent, as Popper seems to suggest. Thus, Simon converts the ambitious aim of searching for a logic of discovery revealing the process of discovery at large, into an unpretentious goal: *'Their modesty [of the examples dealt with] as*

[7]This method is proposed as an alternative to induction. In fact, it made induction absolutely irrelevant; there was no problem of induction as there was any induction procedure as a method for scientific inquiry in the first place, being just an optical illusion to be discarded from the methodology of science.

instances of discovery will be compensated by their transparency in revealing the underlying process' [Sim73a, p. 327].

This humble but brilliant move allows Simon to further draw distinctions on the type of problems to be analysed and on methods to be used. For Simon and his followers, scientific discovery is a problem-solving activity. To this end, a characterization of problems into those that are well structured versus those that are ill structured is provided, and the claim for a logic of discovery focuses mainly on the well-structured ones[8]. A well structured problem is that for which there is a definite criterion for testing, and for which there is at least one problem space in which the initial and the goal state can be represented and all other intermediate states may be reached with appropriate transitions between them. An ill-structured problem lacks at least one of the former conditions.

Although there is no precise methodology by which scientific discovery is achieved, as a form of problem solving, it can be pursued via several methodologies. The key concept in all this is that of *heuristics*, the guide in scientific discovery which is neither totally rational nor absolutely blind. Heuristic methods for discovery are characterized by the use of selective search with fallible results. That is to say, while they provide no complete guarantee to reach a solution, the search in the problem space is not blind, but it is selective according to a predefined strategy. The authors distinguish between 'weak' and 'strong' methods of discovery. The former is the type of problem solving used in novel domains. It is characterized by its generality, since it does not require in-depth knowledge of its particular domain. In contrast, strong methods are used for cases in which our domain knowledge is rich, and are specially designed for one specific structure. Weak methods include generation, testing, heuristic methods, and means-ends analysis, to build explanations and solutions for given problems. These methods have proved useful in artificial intelligence and cognitive simulation, and are used by several computer programs. Examples are the BACON system which simulates the discovery of quantitative laws in Physics (such as Kepler's law and Ohm's law) and the GLAUBER program, which simulates the discovery of qualitative laws in chemistry[9].

Heuristics: A Normative Theory of Discovery

Simon proposes an empirical as well as a normative theory for scientific discovery, both complementing his broad view on this matter:

---

[8]However, cf. [Sim73b] for a proposal on how to treat ill structured problems and for the claim that there is no clear boundary between these kinds of problems.

[9]It is however a matter of debate whether these computer programs really makes discoveries, since it produces theories new to the program but not new to the world, and its discoveries seem spoon–fed rather than created. In the same spirit, Paul Thagard proposes a new field of research, "Computational Philosophy of Science" [Tha88] and puts forward the computational program PI (Processes of Induction) to model some aspects of scientific practice, such as concept formation and theory building.

> *'The theory of scientific discovery has both an empirical part and a formal part. As an empirical theory, it seeks to describe and explain the psychological and sociological processes that are actually employed to make scientific discoveries. As a formal theory, it is concerned with the definition and logical nature of discovery, and it seeks to provide normative advice and prescriptions for anyone who wishes to proceed rationally and efficiently with a task of scientific discovery.'* [Sim77, P. 265]

Let us first deal with the normative aspect of scientific discovery. When Simon proposes a logic for scientific discovery he is relying on a whole research program already developed for artificial intelligence and cognitive psychology, namely the machinery based on heuristic search used for the purposes of devising computer programs to simulate scientific discoveries. For him, a logic of scientific method does not refer to the pure and formal use of the term 'logic', but rather to *'a set of normative standards for judging the processes used to discover or test scientific theories, or the formal structure of the theories themselves. The use of the term 'logic' suggests that the norms can be derived from the goals of the scientific activity. That is to say, a normative theory rests on contingent propositions like: 'If process X is to be efficacious for attaining goal Y, then it should have properties A,B,C'* [Sim73a, p. 328].

The normative theory of scientific discovery is thus a theory for prescribing how actual discoveries may have been made following a rational, though a fallible set of strategies. This view naturally makes his approach close to computational and formal approaches. In fact, Simon views a normative theory of discovery as *'a branch of the theory of computational complexity'* [Sim73a, p. 332]. Moreover, he is willing to use the word 'logic' in a broad sense, something along the lines of a 'rational procedure'. Finally, a normative theory of discovery is one that does not demand a deductive justification for the products of induction, and thus it liberates it from the 'justificationist' positivist view of science.

Concerning the empirical aspect of scientific discovery, we must admit that Simon indeed attempts to describe the psychological processes leading to invent new ideas in science, and in fact talks about incubation and unconscious processes in discovery [Sim77, p. 292], but his position is clearly not aiming to unravel the mysteries of creativity, precisely because he believes there are no such mysteries in the first place: *'new representations, like new problems, do not spring from the brow of Zeus, but emerge by gradual –and very slow– stages ... Even in revolutionary science, which creates those paradigms, the problems and representations are rooted in the past; they are not created out of whole cloth'* [Sim77, p. 301]. Thus, an empirical theory of discovery is based on a psychological theory of discovery, one which provides a mechanistic explanation for thinking, which in turn is demonstrated by programming computers imitating the way humans think. However, on the one hand Simon claims that the processes to explain scientific discovery are only sufficient –and not necessary– to account for discoveries that have actually occurred. On the other hand, he admits there may be forms of making new discoveries in science which may nevertheless be impossible to characterize: *'existing information-processing theories of thinking undoubtedly fall far short of covering the whole range of thinking activities in man. It is not just that the present theories are only approximately correct. Apart from limits on their correctness in explaining the ranges of behaviour to which they apply, much cognitive behaviour* **still** *lies beyond their scope'* [Sim77, p.276]. I leave it as an open question whether this quote, and in general Simon's position, implies that there may be some future time in which an empirical theory of discovery will explain the discovery phenomena at large. I personally believe this is not the case, for he asserts further that creativity in science is achieved by three factors: luck, persistence and 'superior heuristics' (a heuristic is superior to another one if it is a more powerful selective heuristics) [Sim77, p. 290], thus making room to human factors impossible to be characterized in full.

We must note that Simon is not trapped into the discovey–justification dichotomy, but rather views the process of scientific inquiry as a continuum, which only holds the initial discovery and the final justification of a theory as the extreme points of a whole spectrum of reasoning processes. In this respect, his theory is clearly considered as taking a broad view on discovery.

Even though this approach comes from apparently distant disciplines to philosophy of science, namely cognitive psychology and artificial intelligence, they are proposals which suggest the inclusion of computational tools in the philosophy of science research methodology and by so doing claim to reincorporate aspects from the context of discovery within its agenda. However, this approach neither aims nor provides an account of 'the Eureka' moment, even for small ideas.

Although there are several particularities about Simon's proposal with respect to all others, in general his claims reflect the spirit of the project on the search of a logic of discovery for the 'Friends of discovery' overall enterprise.

Science as a Problem - Solving Activity: Popper and Simon Revisited

When presenting the work of these two philosophical giants, we had on focus their views on inquiry in science, in order to explore here to what extent their stances are close together. On the one hand, a first conclusion to draw is that while Popper was genuinely interested in an analysis of new ideas in science, he rendered the very first process of the conception of an idea to be outside the boundaries of the methodology of science, and centered his efforts in giving an account of an ensuing process, that concerned with the methods of analyzing new ideas logically, and accordingly produced his method of conjectures and refutations. Whereas for Simon, his aim was to simulate scientific discovery at large, giving an account both for the generation and evaluation of scientific ideas, convinced that the way to go was to give both an empirical and a normative account of discovery, the former to describe and then represent computationally the intellectual development of human discoveries made in science. The latter to provide prescriptive rules, mainly in the form of heuristic strategies to perform scientific discoveries.

Both authors hold a fallibilist position, one in which there is no certainty of attaining results and where it is possible to refute already assessed knowledge, in favour of new one that better explains the world. However, while for Popper there is one single method for scientific inquiry, the method of conjectures and refutations, for Simon there are several methods for scientific inquiry, for the discovery and justification processes respond to several heuristic strategies, largely based on pattern seeking, the logic of scientific discovery. A further difference between these approaches is found in the method itself for the advancement of science, in what they regard to be the 'logic' for discovery. While for Popper ideas are generated by the method of blind search, Simon and his team develop a full theory to support the view that ideas are generated by the method of 'selective search'. Clearly the latter account allows for a better understanding of how theories and ideas may be generated.

Going back to the three-question division concerning the logic of discovery, namely purpose, pursuit and achievement, it seems useful to analyze these two approaches in regard to them. Popper's purpose for a logic of scientific discovery has no pretensions at all to uncover the epistemics of creativity, but focuses on the evaluation and selection of new ideas in science. Thus, his position is clearly akin with the period corresponding to the search for a logic of justification, as far as the question of purpose is concerned. As to the question of pursuit, his account provides criteria for theory justification under his critical rationalism view, by which no theory is finally settled as true, but is also concerned with the advancement of science, characterizing problem situations as well as a method for their solution. In fact, for Lakatos, Popper's proposal goes beyond its purpose:

> *'There is no infallibilist logic of scientific discovery leading infallibly to results, but there is a fallibilistic logic of discovery which is the logic of scientific progress. But Popper, who has laid the basis for this logic of discovery was not interested in the meta-question of what is the nature of this investigation, so he did not realize that it is neither psychology nor logic, but an independent field, the logic of discovery, heuristics".* [Lak76, p. 167][10].

Therefore, as to the question of achievement, it seems that Popper actually found more than what he was looking for, since his account offered not just

[10] It is interesting to note that in this book, Lakatos was greatly inspired by the history of mathematics, paying particular attention to processes that created new concepts - often referring to G. Polya as the founding father of heuristics in mathematical discovery. (Cf. [Ali00a] for a more detailed discussion of heuristics in the work of G. Polya).

the justification of theories, but in fact a methodology for producing better and stronger theories, something which enters the territory of discovery, at least in so far as the question of the growth of knowledge is concerned.

As for Simon, even though as to the question of purpose he is genuinely after a logic of scientific discovery, his question is also normative, rather than just psychologically prescriptive. And his aim is a modest one, to simulate scientific discoveries for well-structured problems and expressible in some formal or computational language. As to the question of pursuit, by introducing both the notion and the methodology of 'selective search' in a search space for a problem, his research program has achieved the implementation of complex heuristic strategies that simulate well-known scientific discoveries in science. Thus, judging by the empirical examples provided, it is difficult to refute the whole enterprise. Finally, as to the question of achievement, I think Simon's proposal is a coherent one, at least in its normative dimension and modest purpose, it provides results accordingly for a logic of discovery.

To conclude, I think Popper's and Simon's approaches are close together, at least in so far as the following basic ideas are concerned: they both hold a fallibilist stance in regard to the well-foundedness of knowledge and view science as a dynamic activity of problem solving in which the growth of knowledge is the main aspect to characterize, as opposed to the view of science as an static enterprise in search of the assessment of theories as true. But Popper failed to appreciate the philosophical potential of a normative theory of discovery and therefore was blinded to the possibility of devising a logic for the development of knowledge. His view of logic remained static: *'I am quite ready to admit that there is a need for a purely logical analysis of theories, for an analysis which takes no account of how they change and develop. But this kind of analysis does not elucidate those aspects of the empirical science which I, for one, so highly prize'* [Pop59, p. 50].

One reason that allows for the convergence of these two accounts, perhaps obvious by now, is that neither the "Friends of discovery" really account for the epistemics of creativity at large nor Popper neglects its study entirely. Both accounts fall naturally under the study of discovery –when a broad view is endorsed– and neither of them rejects the context of justification, or any other context for that matter. Therefore, it seems that when the focus is on the processes of inquiry in science, rather than on the products themselves, any possible division of contexts of research is doomed to fail sooner or later.

5. Logics for Scientific Methodology

A Place for logic in Scientific Methodology

In this section our aim is to advance the following two claims. In the first place, in our view, the potential for providing a logic for scientific discovery is found in a normative account in the methodology of science. However, we have seen that the formal proposals for this purpose have little, if nothing to do with how a formal logical system looks like. Thus in the second place, our claim is that logic, as understood in modern systems of logic, has a place in the study of logics of discovery. Let us deal with each matter in turn.

The two periods identified above (from antiquity to mid XIXth century) had in mind a logic of discovery of a descriptive nature, one that would capture and describe the way humans reason in science. As noted, these 'logics' had little success, for they failed to provide such an account of discovery. Thereafter, when the search for a logic of discovery was abandoned, a normative account prevailed in favour of proposals of logics of justification. Regarding the approaches of Popper and Simon in this respect, it is clear that while Popper overlooked the possibility of a normative account of logics of discovery, Simon centered his efforts in the development of heuristic procedures, to be implemented computationally, but not on logics –per se– of discovery. In fact, some other proposals which are strictly normative and formal in nature, such as that found in [Ke97], argue for a computational theory as the foundation of a logic of discovery, one which studies algorithmic procedures for the advancement of science.

Therefore, while the way to go seems to be normative, there is nothing in what we have analysed which resembles a logical theory, as understood in the logical tradition. The way the "Friends of Discovery" rightly argue in this respect is to say that the word 'logic' should not at all be interpreted as logic in the traditional sense, and rather think of it as some kind of 'rationality' or 'pattern seeking'. While we believe the approach taken by Simon and his followers has been very fruitful for devising computer programs that simulate scientific discoveries, it has more or less dismissed the place of logic in its endeavors.

Now let us look at the situation nowadays. In spite of few exceptions, logic (classical or otherwise) in philosophy of science is, to put it simply, out of fashion. In fact, although classical logic is part of the curricula in philosophy of science graduate programs, students soon learn that a whole generation of philosophers regarded logical positivism as a failed attempt (though not just for the logic), claiming that scientific practice does not follow logical patterns of reasoning, many of which favoured studies of science based on historical cases. So, why bother about the place of logic in scientific research?

On the one hand, the present situation in logical research has gone far beyond the formal developments that deductive logic reached last century, and new research includes the formalization of several other types or reasoning, like induction and abduction. On the other hand, we claim for a balanced philosophy of science, one in which both methods, the formal and the historical may be complementary, together providing a pluralistic view of science, in which no method is the predominant one. Research nowadays has finally overcome the view that a single treatment in scientific methodology is enough to give us an understanding of scientific practice. A pluralistic view of science was already fostered by Pat Suppes [Sup69] in the sixties.

We know at present that logical models (classical or otherwise) are insufficient to completely characterize notions like explanation, confirmation or falsification in philosophy of science, but this fact does not exclude that some problems in the history of science may be tackled from a formal point of view. For instance, *'some claims about scientific revolutions, seem to require statistical and quantitative data analysis, if there is some serious pretension to regard them with the same status as other claims about social or natural phenomena'* [Sup69, p. 97].

In fact, 'Computational Philosophy of Science' may be regarded as a successful marriage between historical and formal approaches. It is argued that although several heuristic rules have been derived from historical reconstructions in science, they are proposed to be used for future research [MN99]. This is not to say however, that historical analysis of scientific practice could be done in a formal fashion, or that logical treatment should care for some kind of 'historic parameter' in its methodology, but we claim instead that these two views should share their insights and findings in order to complement each other. After all, recall that part of Hempel's inspiration to device a 'logic of explanation' was his inquiry on the logic for 'drawing documents from history'.

Regarding the study of the context of discovery, our position is that it allows for a precise formal treatment. The dominant trend in contemporary philosophy of science neglected the study of the processes of theory discovery in science, partly because other questions were on focus, still we knew nothing about the canons of justification, and the historical approach to scientific practice was making its way into methodological issues. It was research on scientific discovery as problem solving by computational means that really brought back issues of discovery to the philosophy of science agenda. And the field was ready for it, as philosophy of science questions had also evolved. Issues of theory building, concept formation, theory evaluation and scientific progress were at issue. It is now time to unravel these questions with the hand of logical analysis. And here again, we get inspiration, or rather the techniques, from work in logics for common sense reasoning and scientific reasoning developed for artificial intelligence.

Our claim then is that we must bring back logic into scientific methodology, on a par with computational and historical approaches. This move would complement the existing historical account, and as far as its relation to computational approaches in scientific discovery is concerned, it may bring benefits in regard to giving a foundation to the heuristics, to the logic(s) behind the (human) processes of discovery in science. But then, how to integrate a logic in the algorithmic design for scientific discovery? To this end, we must deal with a logic that besides (or even aside from) its semantics and proof theory, gives an account of the search strategy tied to a discovery system. That is, we must provide the automatic procedures to operate a logic, its control strategy, and its procedures to acquire new information without disturbing its coherence and hopefully, achieve some learning in the end.

Logics of Generation and Evaluation

In our view, scientific methodology concerns an inquiry into the methods to account for the dynamics of knowledge. Therefore, the earlier discovery vs. justification dichotomy is transformed into the complementary study of the generation and evaluation of scientific theories. Thus, the pretentious terminology of *logics of discovery* is best labeled as *logics for development* (as argued in [Gut80, p.221]) or *logics of generation and evaluation* , in order to cover both processes of generation and evaluation of theories, and thus vanishing the older dichotomy.

By putting forward a logic for scientific discovery we claim no lack of rigour. But what is clear is that standard deductive logic cannot account for abductive or inductive types of reasoning. And the field is now ready to use new logics for this purpose. An impressive body of results has been obtained about deduction - and a general goal could be the study the wider field of human reasoning while hanging on to these standards of rigour and clarity.

There are still many challenges ahead for the formal study of reasoning in scientific discovery, such as giving an integrated account of deductive, inductive, abductive and analogical styles of inference, the use of diagrams by logical means, and in general the device of logical operations for theory building and change. Already new logical research is moving into these directions. In Burger and Heidema [BH02] degrees of 'abductive boldness' are proposed as spectrum for inferential strength, ranging from cases with poor background information to those with (almost) complete information. Systems dealing with several notions of derivability all at once have also been proposed. A formula may be 'unconditionally derived' or 'conditionally derived', the latter case occurring when a line in a proof asserts a formula which depends on hypotheses which may be later falsified, thus pointing to a notion of proof which allows for addition of lines which are non-deductively derived as well as for deletion of them when falsifying instances occur. This account is found in the framework of (ampliative) adaptive logics, a natural home for abductive inference [Meh05] and for (enumerative) induction inference [Bat05] alike.

The formalization of analogical reasoning is still a growing area of research, without a precise idea of what exactly an analogy amounts to. Perhaps research on mathematical analogy [Pol54], work on analogy in cognitive science [Hel88] or investigations into analogical argumentation theory recently proposed for abduction [GW05], may serve to guide research in this direction. Finally, the study of 'diagrammatic reasoning' is a research field on its own right [BE95], showing that the logical language is not restricted to the two-dimensional left to right syntactic representation, but its agenda still needs to be expanded on research for non-deductive logics.

All the above suggests that the use of non-standard logics to model processes in scientific practice, such as confirmation, falsification, explanation building and theory improvement is, after all, a feasible project. Nevertheless, this claim requires a broad conception of what logic is about.

6. Discussion and Conclusions

In this chapter we claimed that the question of a logic of discovery is better understood by dividing it into three questions, one on its purpose, another one on its pursuit, and finally one on its achievement, for in many proposals there is a clear incongruence between the search and the findings. The search for a logic of discovery was generally (from antiquity to XIXth century) guided by the aim of finding a formal characterization of the ways in which humans reason in scientific discovery, but it has been full of shortcomings, as the products of this search did not succeeded in describing the discovery processes involved in human thought, and were rather directed to the selection and evaluation of already newly discovered ideas in science. Amongst other things, the purely descriptive nature of this enterprise gave place to the position that issues of discovery should be studied psychologically and not philosophically, a view endorsed by Karl Popper under a narrow view of discovery (one which concerns exclusively the initial act of conceiving an idea). However, we showed that under a broad view of discovery (one which conceives processes of discovery at large), Popper's position does include some aspects of discovery having to do with the development of science. In contrast, the view endorsed by Herbert Simon and his followers, claims that there is a normative dimension in the treatment of logics of discovery and taking a finer perspective on the process of discovery as a whole, there is indeed place for proposing several heuristic strategies which characterize and model the generation and evaluation of new ideas in science, at least for well-structured kind of problems.

Therefore, just as we have to acknowledge discovery as a process subject to division, we have also to acknowledge an inescapable part in this process, and accordingly agree with Popper's motto: *'admittedly, creative action can never be fully explained'* [Pop75, p.179]. But this sentence naturally implies that creative action may be partially explained (as argued in [Sav95]). The question then, is to identify a proper line drawing a convenient division in order to carry out an analysis of the generation of new ideas in science.

Our approach in this book in regard to logics of discovery is logical. In regard to the question of purpose we hold a normative account in the methodology of science, one that aims to describe the ways in which a logic of discovery should behave in order to generate new knowledge. As to question of pursuit, we put forward the claim that these logics must find their expression in modern logical systems, as those devised in artificial intelligence.

Chapter 2

WHAT IS ABDUCTION? OVERVIEW AND PROPOSAL FOR INVESTIGATION

1. Introduction

The general purpose of this chapter is to give an overview of the field of abduction in order to provide the conceptual framework of our overall study of abductive reasoning and its relation to explanatory reasoning in subsequent chapters. It is naturally divided into seven parts. After this brief introduction, in the second part (section 2) we motivate our study via several examples that show that this type of reasoning pervades common sense reasoning as well as scientific inquiry. Moreover, abduction may be studied from several perspectives; as a product or as a process, the latter in turn leading to either the process of hypotheses construction or of hypotheses selection and finally, abduction makes sense in connection with its sibling induction, but there are several confusions arising from this relation. In the third part (section 3), we turn to the founding father of abduction, the American pragmatist Charles S. Peirce and present very briefly his theory of abduction. In the fourth part (section 4), we review abduction in the philosophy of science, as it is related with the central topic of scientific explanation, existing both in the received view as well as in neglected ones in this field. In the fifth part (section 5), we present abduction in the field of artificial intelligence and show that it holds a place as a logical inference, as a computational process as well as in theories of belief revision. In the sixth part (section 6), we give an overview of two other fields in which abduction is found, namely in linguistics and in mathematics (neither of which is further pursued in this book). Finally, in the seventh part of this chapter (section 7), we tie up our previous overview by proposing a general taxonomy for abduction, one that allows two different abductive triggers (novelty and anomaly) , which in turn lead to different abductive procedures; and one that allows for several outcomes: facts, rules, or even whole new theories. On our view, abduction is not a new notion of inference. It is rather a topic-dependent practice of scientific reasoning, which can be supported by various types of logical inferences or computational processes.

Each and every part of this overview chapter is further elaborated in subsequent chapters. Thus, the purpose here is just to motivate our study and set the ground for the rest of the book.

2. What is Abduction?

A central theme in the study of human reasoning is the construction of explanations that give us an understanding of the world we live in. Broadly speaking, *abduction* is a reasoning process invoked to explain a puzzling observation. A typical example is a practical competence like medical diagnosis. When a doctor observes a symptom in a patient, she hypothesizes about its possible causes, based on her knowledge of the causal relations between diseases and symptoms. This is a practical setting. Abduction also occurs in more theoretical scientific contexts. For instance, it has been claimed [Han61],[CP, 2.623] that when Kepler discovered that Mars had an elliptical orbit, his reasoning was abductive. But, abduction may also be found in our day-to-day common sense reasoning. If we wake up, and the lawn is wet, we might explain this observation by assuming that it must have rained, or by assuming that the sprinklers had been on. Abduction is thinking from evidence to explanation, a type of reasoning characteristic of many different situations with incomplete information.

The history of this type of reasoning goes back to Antiquity. It has been compared with Aristotle's *apagoge* [CP, 2.776,5.144] which intended to capture a non-strictly deductive type of reasoning whose conclusions are not necessary, but merely possible (not to be confused with *epagoge*, the Aristotelian term for induction). Later on, abduction as reasoning from effects to causes is extensively discussed in Laplace's famous memoirs [Lap04, Sup96] as an important methodology in the sciences. In the modern age, this reasoning was put on the intellectual agenda under the name 'abduction' by C.S. Peirce [CP, 5.189].

To study a type of reasoning that occurs in contexts as varied as scientific discovery, medical diagnosis, and common sense, suitably broad features must be provided, that cover a lot of cases, and yet leave some significant substance to the notion of abduction. The purpose of this preliminary chapter is to introduce these, which will lead to the more specific questions treated in subsequent chapters. But before we start with a more general analysis, let us expand our stock of examples.

Examples

The term 'abduction' is used in the literature for a variety of reasoning processes. We list a few, partly to show what we must cover, and partly, to show what we will leave aside.

1. Common Sense: Explaining observations with simple facts.

All you know is that the lawn gets wet either when it rains, or when the sprinklers are on. You wake up in the morning and notice that the lawn is wet. Therefore you hypothesize that it rained during the night or that the sprinklers had been on.

2. Common Sense: Laying causal connections between facts.

You observe that a certain type of clouds (nimbostratus) usually precede rainfall. You see those clouds from your window at night. Next morning you see that the lawn is wet. Therefore, you infer a causal connection between the nimbostratus at night, and the lawn being wet.

3. Common Sense: Facing a Contradiction.

You know that rain causes the lawn to get wet, and that it is indeed raining. However, you observe that the lawn is not wet. How could you explain this anomaly?

4. Statistical Reasoning: Medical Diagnosis[1].

Jane Jones recovered quite rapidly from a streptococci infection after she was given a dose of penicillin. Almost all streptococcus infections clear up quickly upon administration of penicillin, unless they are penicillin-resistant, in which case the probability of quick recovery is rather small. The doctor knew that Jane's infection is of the penicillin-resistant type, and is completely puzzled by her recovery. Jane Jones then confesses that her grandmother had given her Belladonna, a homeopathic medicine that stimulates the immune system by strengthening the physiological resources of the patient to fight infectious diseases.

The examples so far are fairly typical of what our later analysis can deal with. But actual abductive reasoning can be more complicated than this. For instance, even in common sense settings, there may be various options, which are considered in some sequence, depending on your memory and 'computation strategy'.

5. Common Sense: When something does not work.

[1]This is an adaptation of Hempel's famous illustration of his Inductive-Statistical model of explanation as presented in [Sal92]. The part about homeopathy is entirely mine, however.

You come into your house late at night, and notice that the light in your room, which is always left on, is off. It has being raining very heavily, and so you think some power line went down, but the lights in the rest of the house work fine. Then, you wonder if you left both heaters on, something which usually causes the breakers to cut off, so you check them: but they are OK. Finally, a simpler explanation crosses your mind. Maybe the light bulb of your lamp which you last saw working well, is worn out, and needs replacing.

So, abduction involves computation over various candidates, depending on your background knowledge. In a scientific setting, this means that abductions will depend on the relevant background theory, as well as one's methodological 'working habits'. We mention one often-cited example; even though we should state clearly at this point that it goes far beyond what we shall eventually deal with in our analysis.

6. Scientific Reasoning: Kepler's discovery[2].

One of Johannes Kepler's great discoveries was that the orbit of the planets is elliptical rather than circular. What initially led to this discovery was his observation that the longitudes of Mars did not fit circular orbits. However, before even dreaming that the best explanation involved ellipses instead of circles, he tried several other forms. Moreover, Kepler had to make several other assumptions about the planetary system, without which his discovery does not work. His heliocentric view allowed him to think that the sun, so near to the center of the planetary system, and so large, must somehow cause the planets to move as they do. In addition to this strong conjecture, he also had to generalize his findings for Mars to all planets, by assuming that the same physical conditions obtained throughout the solar system. This whole process of explanation took many years.

It will be clear that the Kepler example has a loose end, so to speak. How we construct the abductive explanation depends on what we take to be his scientific background theory. This is a general feature of abductions: an abductive explanation is always an explanation with respect to some body of beliefs. But even this is not the only parameter that plays a role. One could multiply the above examples, and find still further complicating factors. Sometimes, no single obvious explanation is available, but rather several competing ones - and we have to select. Sometimes, the explanation involves not just advancing facts or rules in our current conceptual frame, but rather the creation of new concepts, that allow for new description of the relevant phenomena. Evidently, we must draw a line somewhere in our present study.

[2]This example is a simplification of one in [Han61].

All our examples were instances of reasoning in which an abductive explanation is needed to account for a certain phenomenon. Is there more unity than this? At first glance, the only clear common feature is that these are not cases of ordinary deductive reasoning, and this for a number of reasons. In particular, the abductive explanations produced might be *defeated.* Maybe the lawn is wet because children have been playing with water. Co-occurrence of clouds and the lawn being wet does not necessarily link them in a causal way. Jane's recovery might after all be due to a normal process of the body. What we learn subsequently can invalidate an earlier abductive conclusion. Moreover, the reasoning involved in these examples seems to go in reverse to ordinary deduction (just as analysis runs in reverse to synthesis, cf. chapter 1), as all these cases run from evidence to hypothesis, and not from data to conclusion, as it is usual in deductive patterns. Finally, describing the way in which an explanation is found, does not seem to follow specific rules. Indeed, the precise nature of Kepler's 'discovery' remains under intensive debate[3].

What we would like to do is the following. Standard deductive logic cannot account for the above types of reasoning. Last century, under the influence of foundational research in mathematics, there has been a contraction of concerns in logic to this deductive core. The result was a loss in breadth, but also a clear gain in depth. By now, an impressive body of results has been obtained about deduction - and we would like to study the wider field of abduction while hanging on to these standards of rigor and clarity. Of course, we cannot do everything at once, and achieve the whole agenda of logic in its traditional open-ended form. We want to find some features of abduction that allow for concrete logical analysis, thereby extending the scope of standard methods. In the next section, we discuss three main features, that occur across all of the above examples (properly viewed), that will be important in our investigation.

Three Faces of Abduction

We shall now introduce three broad oppositions that help in clarifying what abduction is about. At the end, we indicate how these will be dealt with in this book.

[3]For Peirce, Kepler's reasoning was a prime piece of abduction [CP, 1.71,2.96], whereas for Mill it was merely a description of the facts [Mill 58, Bk III, ch II. 3], [CP, 1.71–4]. Even nowadays one finds very different reconstructions. While Hanson presents Kepler's heliocentric view as an essential assumption [Han61], Thagard thinks he could make the discovery assuming instead that the earth was stationary and the sun moves around it [Tha92]. Still a different account of how this discovery can be made is given in [SLB81, LSB87].

Abduction: Product or Process?

The logical key words of *judgment* and *proof* are nouns that denote either an activity, indicated by their corresponding verb, or the result of that activity. In just the same way, the word *abduction* may be used both to refer to a finished product, the *abductive explanation*, or to an activity, the *abductive process* that led to that abductive explanation. These two uses are closely related. An abductive process produces an abductive explanation as its product, but the two are not the same. Note that we can make the same distinction with regard to the notion of *explanation*. We may refer to a finished product, the *explanatory argument*, or to the process of constructing such explanation, the explanatory process.

One can relate these distinctions to more traditional ones. An example is the earlier opposition of 'context of discovery' versus 'context of justification' (cf. chapter 1). Kepler's abductive explanation–product "the orbit of the planets is elliptical", which justifies the observed facts, does not include the abductive–process of how he came to make this discovery. The context of discovery has often been taken to be purely psychological, but this does not preclude its exact study. Cognitive psychologists study mental patterns of discovery, learning theorists in AI study formal hypothesis formation, and one can even work with concrete computational algorithms that produce abductive explanations. To be sure, it is a matter of debate whether Kepler's reasoning may be modeled by a computer. (For a computer program that claims to model this particular discovery, cf. [SLB81].) However this may be, one can certainly write simple programs that produce common sense explanations of 'why the lawn is wet', as we will show later on.

Moreover, once produced, abductive explanations are public objects of "justification", which can be checked and tested by independent logical criteria. An abductive explanation is an element in an *explanatory argument*, which is in turn the product of an *explanatory process*. So, just as abduction, explanation has also its product and its process sides. The overall procedure as to how these two reasoning processes can be distinguished and how they are connected to each other can be exemplified as follows: We may distinguish between the process of the discovery of a geometrical proof, which produces the required postulates to be used as premisses and possibly some auxiliary constructions, and between the actual process of proving the desired theorem, which in turn produces a proof as a logical argument.

The product–process distinction has been recognized by logicians [Bet59, vBe93], in the context of deductive reasoning, as well as by philosophers of science [Rub90, Sal90] in the context of scientific explanation. Both lead to interesting questions by themselves, and so does their interplay. Likewise, these two faces of abduction are both relevant for our study. On the product side, our focus will be on conditions that give a piece of information explanatory force, and on the process side, we will be concerned with the design of algorithms that produce abductive explanations.

Abduction: Construction or Selection?

Given a fact to be explained, there are often several possible abductive explanations, but only one (or a few) that counts as the best one. Pending subsequent testing, in our common sense example of light failure, several abductive explanations account for the unexpected darkness of the room (power line down, breakers cut off, bulb worn out). But only one may be considered as 'best' explaining the event, namely the one that really happened. But other preference criteria may be appropriate, too, especially when we have no direct test available.

Thus, abduction is connected to both hypothesis construction and hypothesis selection. Some authors consider these processes as two separate steps, construction dealing with what counts as a possible abductive explanation, and selection with applying some preference criterion over possible abductive explanations to select the best one. Other authors regard abduction as a single process by which a single best explanation is constructed. Our position is an intermediate one. We will split abduction into a first phase of hypothesis construction, but also acknowledge a next phase of hypothesis selection. We shall mainly focus on a characterization of possible abductive explanations. We will argue that the notion of a 'best abductive explanation' necessarily involves contextual aspects, varying from application to application. There is at least a new parameter of preference ranking here. Although there exist both a philosophical tradition on the logic of preference [Wri63], and logical systems in AI for handling preferences that may be used to single out best explanations [Sho88, DP91b], the resulting study would take us too far afield.

Abduction vs. Induction

Once beyond deductive logic, diverse terminologies are being used. Perhaps the most widely used term is inductive reasoning [Mill 58, Sal90, HHN86, Tha88, Fla95, Mic94]. Abduction is another focus, and it is important, at least, to clarify its relationship to induction. For C.S. Peirce, as we shall see, 'deduction', 'induction' and 'abduction' formed a natural triangle – but the literature in general shows many overlaps, and even confusions.

Since the time of John Stuart Mill (1806-1873), the technical name given to all kinds of non-deductive reasoning has been 'induction', though several *methods for discovery and demonstration of causal relationships* [Mill 58] were recognized. These included generalizing from a sample to a general property, and reasoning from data to a causal hypothesis (the latter further divided into methods of agreement, difference, residues, and concomitant variation). A more refined and modern terminology is 'enumerative induction' and 'explanatory induction', of which 'inductive generalization', 'inductive projection', 'statistical syllogism', 'concept formation' are some instances. Such a broad connotation of the term induction continues to the present day. For instance, in the computational philosophy of science, induction is understood *"in the broad sense of any kind of inference that expands knowledge in the face of uncertainty"* [Tha88].

Another 'heavy term' for non-deductive reasoning is *statistical reasoning*, introducing a probabilistic flavour, like our example of medical diagnosis, in which possible explanations are not certain but only probable. Statistical reasoning exhibits the same diversity as abduction. First of all, just as the latter is strongly identified with *backwards deduction* (as we shall see later on in this chapter), the former finds its 'reverse notion' in probability[4]. Both abduction and statistical reasoning are closely linked with notions like confirmation (the testing of hypothesis) and likelihood (a measure for alternative hypotheses).

On the other hand, some authors take induction as an instance of abduction. Abduction as *inference to the best explanation* is considered by Harman [Har65] as the basic form of non-deductive inference, which includes (enumerative) induction as a special case.

This confusion returns in artificial intelligence. 'Induction' is used for the process of learning from examples – but also for creating a theory to explain the observed facts [Sha91], thus making abduction an instance of induction. Abduction is usually restricted to producing abductive explanations in the form of facts. When the explanations are rules, it is regarded as part of induction. The relationship between abduction and induction (properly defined) has been the topic for workshops in AI conferences [ECAI96] and edited books [FK00].

To clear up all these conflicts, one might want to coin new terminology altogether. Many authors write as if there were pre-ordained, reasonably clear notions of abduction and its rivals, which we only have to analyze to get a clear picture. But these technical terms may be irretrievably confused in their full generality, burdened with the debris of defunct philosophical theories. Therefore, I have argued for a new term of *"explanatory reasoning"* in [Ali96a], trying to describe its fundamental aspects without having to decide if they are instances of either abduction or induction. In this broader perspective, we can also capture explanation for more than one instance or for generalizations, – which we have not mentioned at all – and introduce further fine–structure. For example, given two observed events, in order to find an explanation that accounts for them, it must be decided whether they are causally connected (eg. entering the copier room and the lights going on), correlated with a common cause (e.g. observing both the barometric pressure and the temperature drop-

[4]The problem in probability is: given an stochastic model, what can we say about the outcomes? The problem in statistics is the reverse: given a set of outcomes, what can we say about the model?.

ping at the same time), or just coincidental without any link (you reading this paragraph in place A while I revise it somewhere in place B). But in this book, we shall concentrate on explanatory reasoning from simple facts, giving us enough variety for now. Hopefully, this case study of abduction will lead to broader clarity of definition as well.

More precisely, we shall understand abduction as reasoning from *a single observation to its abductive explanations*, and induction as *enumerative induction* from samples to general statements. While induction explains a set of observations, abduction explains a single one. Induction makes a prediction for further observations, abduction does not (directly) account for later observations. While induction needs no background theory per se, abduction relies on a background theory to construct and test its abductive explanations. (Note that this abductive formulation does not commit ourselves to any specific logical inference, kind of observation, or form of explanation.)

As for their similarities, induction and abduction are both non-monotonic types of inference (to be defined in chapter 3), and both run in opposite direction to standard deduction. In non-monotonic inference, new premises may invalidate a previous valid argument. In the terminology of philosophers of science, non-monotonic inferences are not *erosion proof* [Sal92]. Qua direction, induction and abduction both run from evidence to explanation. In logical terms, this may be viewed as going from a conclusion to (part of) its premises, in reverse of ordinary deduction. We will return to these issues in much more detail in our logical chapter 3.

3. The Founding Father: C.S. Peirce

The literature on abduction is so vast, that we cannot undertake a complete survey here. What we shall do is survey some highlights, starting with the historical sources of the modern use of the term. In this field, all roads from XXth century onwards lead back to the work of C.S. Peirce. Together with the other sources to be discussed, the coming sections will lead up to further parameters for the general taxonomy of abduction that we propose toward the end of this chapter.

Understanding Peirce's Abduction

Charles Sanders Peirce (1839-1914), the founder of American pragmatism was the first philosopher to give to abduction a logical form, and hence his relevance to our study. However, his notion of abduction is a difficult one to unravel. On the one hand, it is entangled with many other aspects of his philosophy, and on the other hand, several different conceptions of abduction evolved in his thought. The notions of logical inference and of validity that Peirce puts forward go beyond our present understanding of what logic is about. They are linked to his epistemology, a dynamic view of thought as logical inquiry, and correspond to a deep philosophical concern, that of studying the nature of synthetic reasoning.

We will point out a few general aspects of his later theory of abduction, and then concentrate on some of its more logical aspects. For a more elaborate analysis of abduction in connection to Peirce's epistemology and pragmatism, cf. chapter 7.

The Key Features of Peircean Abduction

For Peirce, three aspects determine whether a hypothesis is promising: it must be *explanatory*, *testable*, and *economic*. A hypothesis is an explanation if it accounts for the facts. Its status is that of a suggestion until it is verified, which explains the need for the testability criterion.

Finally, the motivation for the economic criterion is twofold: a response to the practical problem of having innumerable explanatory hypotheses to test, as well as the need for a criterion to select the best explanation amongst the testable ones.

For the explanatory aspect, Pierce gave the following often-quoted logical formulation [CP, 5.189]:

The surprising fact, C, is observed.

But if A were true, C would be a matter of course.

Hence, there is reason to suspect that A is true.

This formulation has played a key role in Peirce scholarship, and it has been the point of departure of many classic studies on abductive reasoning in artificial intelligence [FK00], such as in logic programming [KKT95], knowledge acquisition [KM90] and natural language processing [HSAM90]. Nevertheless, these approaches have paid little attention to the elements of this formulation and none to what Peirce said elsewhere in his writings. This situation may be due to the fact that his philosophy is very complex and not easy to be implemented in the computational realm. The notions of logical inference and of validity that Peirce puts forward go beyond logical formulations. In our view, however, there are several aspects of Peirce's abduction which are tractable and may be implemented using machinery of AI, such as that found in theories of belief revision (cf. chapter 8).

Our own understanding of abductive reasoning reflects this Peircean diversity in part, taking abduction as a style of logical reasoning that occurs at different levels and in several degrees. These will be reflected in our proposal for a taxonomy with several 'parameters' for abductive reasoning.

In AI circles, this formulation has been generally interpreted as the following logical argument–schema:

C

$\underline{A \rightarrow C}$

A

Where the status of A is tentative (it does not follow as a logical consequence from the premises).

However intuitive, this interpretation certainly captures neither the fact that C is surprising nor the additional criteria Peirce proposed. Moreover, the interpretation of the second premise is not committed to material implication. In fact, some have argued (cf. [FK00]) that this is a vacuous interpretation and favour one of classical logical entailment ($A \models C$). But other interpretations are possible; any other nonstandard form of logical entailment or even a computational process in which A is the input and C the output, are all feasible interpretations for "if C were true, A would be a matter of course".

The additional Peircean requirements of testability and economy are not recognized as such in AI, but are nevertheless incorporated. The latter criterion is implemented as a further selection process to produce the *best explanation*, since there might be several formulae that satisfy the above formulation but are not appropriate as explanations. As for the testability requirement, when the second premise is interpreted as logical entailment this requirement is trivialized, since given that C is true, in the simplest sense of 'testable', A will always be testable.

We leave here the reconstruction of Peirce's notion of abduction. For further aspects, we refer to chapter 7. The additional criteria of testability and economy are not part of our general framework for abduction. Testability as understood by Peirce is an extra-logical empirical criterion, while economy concerns the selection of explanations, which we already put aside as a further stage of abduction requiring a separate study.

4. Philosophy of Science

Peirce's work stands at the crossroads of many traditions, including logic, epistemology, and philosophy of science. Especially, the latter field has continued many of his central concerns. Abduction is clearly akin to core notions of modern methodology, such as explanation, induction, discovery, and heuristics. We have already discussed a connection between process-product aspects of both abduction and explanation and the well-known division between contexts of discovery and justification. We shall discuss several further points of contact in chapters 5 and 6 below. But for the moment, we start with a first foray.

The 'Received View'

The dominant trend in philosophy has focused on abduction and explanation as products rather than a process, just as it has done for other epistemic notions. Aristotle, Mill, and in this century, the influential philosopher of science Carl Hempel, all based their accounts of explanation on proposing criteria to characterize its products. These accounts generally classify into argumentative and non-argumentative types of explanation [Rub90, Sal90, Nag79]. Of particular importance to us is the 'argumentative' Hempelian tradition. Its followers aim to model empirical why-questions, whose answers are scientific explanations in the form of arguments. In these arguments, the 'explanandum' (the fact to be explained) is derived (deductively or inductively) from the 'explananda' (that which does the explaining) supplemented with relevant 'laws' (general or statistical) and 'initial conditions'. For instance, the fact that an explosion occurred may be explained by my lighting the match, given the laws of physics, and initial conditions to the effect that oxygen was present, the match was not wet, etcetera.

In its deductive version, the Hempelian account, found in many standard texts on the philosophy of science [Nag79, Sal90] is called *deductive-nomological*, for obvious reasons. But its engine is not just standard deduction. Additional restrictions must be imposed on the relation between explananda and explanandum, as neither deduction nor induction is a sufficient condition for genuine explanation. To mention a simple example, every formula is derivable from itself ($\varphi \vdash \varphi$), but it seems counterintuitive, or at least very uninformative, to explain anything by itself.

Other, non-deductive approaches to explanation exist in the literature. For instance, [Rub90] points at these two:

[Sal77, p.159] takes them to be: *"an assemblage of factors that are statistically relevant"*.

While [vFr80, p.134] makes them simply: *"an answer"*.

For Salmon, the question is not how probable the explanans renders the explanandum, but rather whether the facts adduced make a difference to the probability of the explanandum. Moreover, this relationship need not be in the form of an argument. For van Fraassen, a representative of pragmatic approaches to explanation, the explanandum is a contrastive why-question. Thus, rather than asking "why φ?", one asks "why φ rather than γ?". The pragmatic view seems closer to abduction as a process, and indeed, the focus on questions introduces some dynamics of explaining. Still, it does not tell us how to produce explanations.

There are also alternative deductive approaches. An example is the work of Rescher [Res78], which introduces a *direction of thought*. Interestingly, this establishes a temporal distinction between 'prediction' and 'retroduction' (Rescher's term for abduction), by marking the precedence of the explanandum over the hypothesis in the latter case. Another, and quite famous deductivist tradition is Popper's logic of scientific discovery [Pop59], which we already analyzed in chapter 1. Its method of conjectures and refutations proposes the testing of hypotheses, by attempting to refute them.

What is common to all these approaches in the philosophy of science is the importance of a hidden parameter in abduction. Whether with Hempel, Salmon, or Popper, scientific explanation never takes place in isolation, but always in the context of some *background theory*. This additional parameter will become part of our general scheme to be proposed below.

The 'Neglected View'

Much more marginal in the philosophy of science are accounts of abduction and explanation that focus on the processes as such. One early author emphasizing explanation as a process of discovery is Hanson ([Han61]), who gave an account of patterns of discovery, recognizing a central role for abduction (which he calls 'retroduction'). Also relevant here is the work by Lakatos ([Lak76]), a critical response to Popper's logic of scientific discovery. As already noted in chapter 1, For Lakatos, there is only a fallibilistic logic of discovery, which is neither psychology nor logic, but an independent discipline, the *logic of heuristics*. He pays particular attention to processes that created new concepts in mathematics – often referring to Polya ([Pol45]) as the founding father of heuristics in mathematical discovery[5]. We will come back to this issue later in the chapter, when presenting further fields of application.

What these examples reveal is that in science, explanation involves the invention of new concepts, just as much as the positing of new statements (in some fixed conceptual framework). So far, this has not led to extensive formal studies of concept formation, similar to what is known about deductive logic. (Exceptions that prove the rule are occasional uses of Beth's Definability Theorem in the philosophical literature. A similar lacuna vis-a-vis concept revision exists in the current theory of belief revision in AI, cf. chapter 8.). We are certainly in sympathy with the demand for conceptual change in explanation – but this topic will remain beyond the technical scope of this book (however, cf. chapter 4 for a brief discussion).

5. Artificial Intelligence

Our next area of comparison is artificial intelligence. The transition with the previous section is less abrupt than it may be seem. It has often been noted, by looking at the respective research agendas, that artificial intelligence

[5]In fact Polya contrasts two types of arguments. A demonstrative syllogism in which from $A \rightarrow B$, and B false, $\neg A$ is concluded, and a *heuristic syllogism* in which from $A \rightarrow B$, and B true, it follows that A is more credible. The latter, of course, recalls Peirce's abductive formulation.

is philosophy of science, pursued by other means (cf. [Tan92]). Research on abductive reasoning in AI dates back to 1973 [Pop73], but it is only fairly recently that it has attracted great interest, in areas like logic programming [KKT95], knowledge assimilation [KM90], and diagnosis [PG87], to name just a few. Abduction is also coming up in the context of data bases and knowledge bases, that is, in mainstream computer science.

In this setting, the product-process distinction has a natural counterpart, namely, in logic-based vs computational-based approaches to abduction. While the former focuses on giving a semantics to the logic of abduction, usually defined as 'backwards deduction plus additional conditions', the latter is concerned with providing algorithms to produce abductions.

It is impossible to give an overview here of this exploding field. Therefore, we limit ourselves to (1) a brief description of abduction as logical inference, (2) a presentation of abduction in logic programming, and (3) a sketch of the relevance of abductive thinking in knowledge representation. There is much more in this field of potential philosophical interest, however. For abduction in bayesian networks, connectionism, and many other angles, the reader is advised to consult [Jos94, PR90, Kon96, Pau93, AAA90, FK00].

Abduction as Logical Inference

The general trend in logic based approaches to abduction in AI interprets abduction as *backwards deduction plus additional conditions*. The idea of abduction as deduction in reverse plus additional conditions brings it very close to deductive-nomological explanation in the Hempel style (cf. Chapter 5), witness the following format. What follows is the 'standard version' of abduction as deduction via some consistent additional assumption, satisfying certain extra conditions. It combines some common requirements from the literature (cf. [Kon90, KKT95, MP93] and chapter 3 for further motivation). Note however that given the distinction (product vs. process) and the terminology introduced so far (abductive explanation, explanatory argument), the following definition is one which characterizes the conditions for an abductive explanation to be part of an explanatory argument. Thus this is a characterization of the forward inference from theory and abduced hypothesis to evidence (rather than the backward inference –which is indeed what mimics the abductive process– triggered by an evidence taking place with respect to a background theory and producing an abductive explanation), and we will referred to it as an *(abductive) explanatory argument*, in order to highlight that the argument is explanatory, while keeping in mind it aims to characterize the conditions for an abductive explanation:

(Abductive) Explanatory Argument:

Given a theory Θ (a set of formulae) and a formula φ (an atomic formula), α is an *abductive explanation* if

1 $\Theta \cup \alpha \models \varphi$

2 α is consistent with Θ

3 α is 'minimal' (there are several ways to characterize minimality, to be discussed in chapter 3).

4 α has some restricted syntactical form (usually an atomic formula or a conjunction of them).

An additional condition not always made explicit is that $\Theta \not\models \varphi$. This says that the fact to be explained should not already follow from the background theory alone. Sometimes, the latter condition figures as a precondition for an *abductive problem*.

What can one say in general about the properties of such an 'enriched' notion of consequence? As we have mentioned before, a new logical trend in AI studies variations of classical consequence via their 'structural rules', which govern the combination of basic inferences, without referring to any special logical connectives. (Cf. the analysis of non-monotonic consequence relations in AI of [Gab96], [KLM90], and the analysis of dynamic styles of inference in linguistics and cognition in [vBe90].) Perhaps the first example of this approach in abduction is the work in [Fla95] – and indeed our analysis in chapter 3 will follow this same pattern.

Abduction in Logic Programming

Logic Programming [LLo87, Kow79] was introduced by Bob Kowalski and Alan Colmerauer in 1974, and is implemented as (amongst others) the programming language Prolog. It is inspired by first-order logic, and it consists of logic programs, queries, and a underlying inferential mechanism known as resolution[6].

Abduction emerges naturally in logic programming as a 'repair mechanism', completing a program with the facts needed for a query to succeed. This may be illustrated by our rain example (1) from the introduction in Prolog:

Program P:

lawn-wet $\leftarrow$ rain.

lawn-wet $\leftarrow$ sprinklers-on.

Query q: lawn-wet.

[6]Roughly speaking, a Prolog program P is an ordered set of rules and facts. Rules are restricted to horn-clause form $A \leftarrow L_1, \ldots, L_n$ in which each L_i is an atom A_i. A query q (theorem) is posed to program P to be solved (proved). If the query follows from the program, a positive answer is produced, and so the query is said to be successful. Otherwise, a negative answer is produced, indicating that the query has failed. However, the interpretation of negation is 'by failure'. That is, 'no' means 'it is not derivable from the available information in P' – without implying that the negation of the query $\neg q$ is derivable instead. Resolution is an inferential mechanism based on refutation working backwards: from the negation of the query to the data in the program. In the course of this process, valuable by-products appear: the so-called 'computed answer substitutions', which give more detailed information on the objects satisfying given queries.

Given program P, query q does not succeed because it is not derivable from the program. For q to succeed, either one (or all) of the facts 'rain', 'sprinklers-on', 'lawn-wet' would have to be added to the program. Abduction is the process by which these additional facts are produced. This is done via an extension of the resolution mechanism that comes into play when the backtracking mechanism fails. In our example above, instead of declaring failure when either of the above facts is not found in the program, they are marked as 'hypothesis' (our abductive explanations), and proposed as those formulas which, if added to the program, would make the query succeed.

In actual *Abductive Logic Programming* [KKT95], for these facts to be counted as abductions, they have to belong to a pre-defined set of 'abducibles', and to be verified by additional conditions (so-called 'integrity constraints'), in order to prevent a combinatorial explosion of possible explanations.

In logic programming, the procedure for constructing explanations is left entirely to the resolution mechanism, which affects not only the order in which the possible explanations are produced, but also restricts the form of explanations. Notice that rules cannot occur as abducibles, since explanations are produced out of sub-goal literals that fail during the backtracking mechanism. Therefore, our common sense example (2) in which a causal connection is abduced to explain why the lawn is wet, cannot be implemented in logic programming[7]. The additional restrictions select the best hypothesis. Thus, processes of both construction and selection of explanations are clearly marked in logic programming[8].

Logic programming does not use blind deduction. Different control mechanisms for proof search determine how queries are processed. This additional degree of freedom is crucial to the efficiency of the enterprise. Hence, different control policies will vary in the abductions produced, their form and the order in which they appear. To us, this variety suggests that the procedural notion of abduction is intensional, and must be identified with different practices, rather than with one deterministic fixed procedure.

Abduction and Theories of Epistemic Change

Most of the logic-based and computation-based approaches to abduction reviewed in the preceding sections assume that neither the explanandum nor its negation is derivable from the background theory ($\Theta \not\models \varphi$, $\Theta \not\models \neg\varphi$). This leaves no room to represent problems like our common sense light example (5) in which the theory expects the contrary of our observation. (Namely, that the

[7]At least, this is how the implementation of abduction in logic programming stands as of now. It is of course possible to write extended programs that produce these type of explanations.

[8]Another relevant connection here is research in 'inductive logic programming' ([Mic94], [FK00]) which integrates abduction and induction.

light in my room is on.) These are cases where the theory needs to be *revised* in order to account for the observation. Such cases arise in practical settings like diagnostic reasoning [PR90], belief revision in databases [AD94] and theory refinement in machine learning [SL90, Gin88].

When importing revision into abductive reasoning, an obvious related territory is theories of belief change in AI. Mostly inspired by the work of Gärdenfors [Gar88] (a work whose roots lie in the philosophy of science), these theories describe how to incorporate a new piece of information into a database, a scientific theory, or a set of common sense beliefs. The three main types of belief change are operations of 'expansion', 'contraction', and 'revision'. A theory may be expanded by adding new formulas, contracted by deleting existing formulas, or revised by first being contracted and then expanded. These operations are defined in such a way as to ensure that the theory or belief system remains consistent and suitably 'closed' when incorporating new information.

Our earlier cases of abduction may be described now as expansions, where the background theory gets extended to account for a new fact. What is added are cases where the surprising observation (in Peirce's sense) calls for revision. Either way, this perspective highlights the essential role of the background theory in explanation. Belief revision theories provide an explicit calculus of modification for both cases. It must be clarified however, that changes occur only in the theory, as the situation or world to be modeled is supposed to be static, only new information is coming in. Another important type of epistemic change studied in AI is that of *update*, the process of keeping beliefs up–to–date as the world changes. We will not analyze this second process here – even though we are confident that it can be done in the same style proposed here. Evidently, all this ties in very neatly with our earlier findings. (For instance, the theories involved in abductive belief revision might be structured like those provided by our discussion, or by cues from the philosophy of science.) We will explore this connection in more detail in chapter 8.

6. Further Fields of Application

The above survey is by no means exhaustive. Abduction occurs in many other research areas, of which we will mention two: linguistics and mathematics. Although we will not pursue these lines elsewhere in this book, they do provide a better perspective on the broader interest of our topic.

Abduction in Linguistics

In linguistics, abduction has been proposed as a process for natural language interpretation [HSAM90], where our 'observations' are the utterances that we hear (or read). More precisely, interpreting a sentence in discourse is viewed as providing a best explanation of why the sentence would be true. For instance, a listener or reader abduces assumptions in order to resolve references for definite descriptions ("the cat is on the mat" invites you to assume that there is a cat and a mat), and dynamically accommodates them as presupposed information for the sentence being heard or read.

Abduction also finds a place in theories of language acquisition. Most prominently, Chomsky proposed that learning a language is a process of theory construction. A child 'abduces' the rules of grammar guided by her innate knowledge of language universals. Indeed in [Cho72], he refers to Peirce's justification for the logic of abduction, – based on the human capacity for 'guessing the right hypotheses', to reinforce his claim that language acquisition from highly restricted data is possible.

Abduction has also been used in the semantics of *questions*. Questions are then the input to abductive procedures generating answers to them. Some work has been done in this direction in natural language as a mechanism for dealing with indirect replies to yes-no questions [GJK94]. Of course, the most obvious case where abduction is explicitly called for are "Why" questions, inviting the other person to provide a reason or cause.

Abduction also connects up with linguistic *presuppositions*, which are progressively accommodated during a discourse. The phenomenon of accommodation is a non-monotonic process, in which presuppositions are not direct updates for explicit utterances, but rather abductions that can be refuted because of later information. Accommodation can be described as a *repair strategy* in which the presuppositions to be accommodated are not part of the background. In fact, the linguistic literature has finer views of types of accommodation (cf. the 'local'/'global' distinction in [Hei83]), which might correspond to the two abductive 'triggers' proposed in the next section. A broader study on presuppositions which considers abductive mechanisms and uses the framework of semantic tableaux to represent the context of discourse, is found in [Ger95].

More generally, we feel that the taxonomy proposed later in this chapter might correlate with the linguistic diversity of presupposition (triggered by definite descriptions, verbs, adverbs, et cetera) – but we must leave this as an open question.

Abduction in Mathematics

Abduction in mathematics is usually identified with notions like discovery and heuristics. A key reference in this area is the work by the earlier mentioned G. Polya [Pol45, Pol54, Pol62]. In the context of number theory, for example, a general property may be guessed by observing some relation as in:

$$3+7=10, \qquad 3+17=20, \qquad 13+17=30$$

Notice that the numbers 3,7,13,17 are all odd primes and that the sum of any of two of them is an even number. An initial observation of this kind eventually led

Goldbach (with the help of Euler) to formulate his famous conjecture: *'Every even number greater than two is the sum of two odd primes'*.

Another example is found in a configuration of numbers, such as in the well–known Pascal's triangle [Pol62]:

$$
\begin{array}{ccccccccc}
 & & & 1 & & & & = & 1 \\
 & & 1 & & 1 & & & = & 2 \\
 & 1 & & 2 & & 1 & & = & 4 \\
1 & & 3 & & 3 & & 1 & = & 8
\end{array}
$$

There are several 'hidden' properties in this triangle, which the reader may or may not discover depending on hcr previous training and mathematical knowledge. A simple one is that any number different from 1 is the sum of two other numbers in the array, namely of its northwest and northeast neighbors (e.g. 3 = 1 + 2). A more complex relationship is this: the numbers in any row sum to a power of 2. More precisely,

$$\binom{n}{0} + \dots + \binom{n}{n} = 2^n$$

See [Pol62] for more details on 'abducing' laws about the binomial coefficients in the Pascal Triangle.

The next step is to ask why these properties hold, and then proceed to prove them. Goldbach's conjecture remains unsolved (it has not been possible to prove it or to refute it); it has only been verified for a large number of cases (the latest news is that it is true for all integers less than 4.10^{11}, cf. [Rib96]). The results regarding Pascal's triangle on the other hand, have many different proofs, depending one's particular taste and knowledge of geometrical, recursive, and computational methods. (Cf. [Pol62] for a detailed discussion of alternative proofs.)

According to Polya, a mathematician discovers just as a naturalist does, by observing the collection of his specimens (numbers or birds) and then guessing their connection and relationship [Pol54, p.47]. However, the two differ in that verification by observation for the naturalist is enough, whereas the mathematician requires a proof to accept her findings. This points to a unique feature of mathematics: once a theorem finds a proof, it cannot be defeated. Thus, mathematical truths are eternal, with possibly many ways of being *explained*. On the other hand, some findings may remain unexplained forever. Abduction in mathematics shows very well that observing goes beyond visual perception, as familiarity with the field is required to find 'surprising facts'. Moreover, the relationship between observation and proof need not be causal, it is just pure mathematical structure that links them together.

Much more complex cases of mathematical discovery can be studied, in which concept formationis involved. An interesting approach along these lines is found in [Vis97], which proposes a catalogue of procedures for creating concepts when solving problems. These include 'redescription', 'substitution', and 'transposition', which are explicitly related to Peirce's treatment of abduction.

7. A Taxonomy for Abduction

What we have seen so far may be summarized as follows. Abduction is a process whose products are specific abductive explanations, with a certain inferential structure, making an (abductive) explanatory argument. We consider the two aspects of abduction of equal importance. Moreover, on the process side, we distinguished between constructing possible abductive explanations and selecting the best one amongst these. This book is mainly concerned with the characterization of abductive explanations as products –as an essential ingredient for explanatory arguments– and with the processes for constructing these.

As for the *logical form* of abduction –referring to the inference corresponding to the abductive process that takes a background theory (Θ) and a given observation (φ) as inputs, and produces an abductive explanation (α) as its output– we have found that at a very general level, it may be viewed as a threefold relation:

$$\Theta, \varphi \Rightarrow \alpha$$

Other parameters are possible here, such as a preference ranking - but these would rather concern the further selection process. This characterization aims to capture the direction (from evidence to abductive explanation) of this type of reasoning. In the end however, what we want is to characterize an (abductive) explanatory argument, in its deductive forward fashion, that is, an inference from theory (Θ) and abductive explanation (α) to evidence (φ) as follows:

$$\Theta, \alpha \Rightarrow \varphi$$

Against this background, we propose three main parameters that determine types of explanatory arguments. (i) An 'inferential parameter' ($\Rightarrow$) sets some suitable logical relationship among explananda, background theory, and explanandum. (ii) Next, 'triggers' determine what kind of abductive process is to be performed: φ may be a novel phenomenon, or it may be in conflict with the theory Θ. (iii) Finally, 'outcomes' (α) are the various products (abductive explanations) of an abductive process: facts, rules, or even new theories.

Abductive Parameters

Varieties of Inference

In the above schema, the notion of explanatory inference $\Rightarrow$ is not fixed. It can be classical derivability $\vdash$ or semantic entailment $\models$, but it does not have to be. Instead, we regard it as a parameter which can be set independently. It ranges over such diverse values as probable inference ($\Theta, \alpha \Rightarrow_{probable} \varphi$), in which the explanans renders the explanandum only highly probable, or as the inferential mechanism of logic programming ($\Theta, \alpha \Rightarrow_{prolog} \varphi$). Further interpretations include dynamic inference ($\Theta, \alpha \Rightarrow_{dynamic} \varphi$, cf. [vBe96a]), replacing truth by information change potential along the lines of belief update or revision. Our point here is that abduction is not one specific non-standard logical inference mechanism, but rather a way of using any one of these.

Different Triggers

According to Peirce, as we saw, abductive reasoning is triggered by a *surprising phenomenon.* The notion of surprise, however, is a relative one, for a fact φ is surprising only with respect to some background theory Θ providing 'expectations'. What is surprising to me (eg. that the lights go on as I enter the copier room) might not be surprising to you. We interpret a surprising fact as one which needs an explanation. From a logical point of view, this assumes that the fact is not already explained by the background theory Θ: $\Theta \not\Rightarrow \varphi$.

Moreover, our claim is that one also needs to consider the status of the negation of φ. Does the theory explain the negation of observation instead ($\Theta \Rightarrow \neg\varphi$)? Thus, we identify at least two triggers for abduction: *novelty* and *anomaly*:

- Abductive Novelty: $\Theta \not\Rightarrow \varphi, \Theta \not\Rightarrow \neg\varphi$

  φ is novel. It cannot be explained ($\Theta \not\Rightarrow \varphi$), but it is consistent with the theory ($\Theta \not\Rightarrow \neg\varphi$).

- Abductive Anomaly: $\Theta \not\Rightarrow \varphi, \Theta \Rightarrow \neg\varphi$.

  φ is anomalous. The theory explains rather its negation ($\Theta \Rightarrow \neg\varphi$).

In the computational literature on abduction, novelty is the condition for an abductive, problem*abductive problem* [KKT95]. My suggestion is to incorporate anomaly as a second basic type.

Of course, non-surprising facts (where $\Theta \Rightarrow \varphi$) should not be candidates for abductive explanations. Even so, one might speculate if facts which are merely probable on the basis of Θ might still need an abductive explanation of some sort to further cement their status.

Different Outcomes

Abductive explanations themselves come in various forms: facts, rules, or even theories. Sometimes one simple fact suffices to explain a surprising phenomenon, such as rain explaining why the lawn is wet. In other cases, a rule establishing a causal connection might serve as an abductive explanation, as in our case connecting cloud types with rainfall. And many cases of abduction in science provide new theories to explain surprising facts. These different options may sometimes exist for the same observation, depending on how seriously we want to take it. In this book, we shall mainly consider abductive explanations in the forms of atomic facts, conjunctions of them and simple conditionals, but we do make occasional excursions to more complex kinds of statements.

Moreover, we are aware of the fact that genuine abductive explanations sometimes introduce new concepts, over and above the given vocabulary. (For instance, the eventual explanation of planetary motion was not Kepler's, but Newton's, who introduced a new notion of 'force' – and then derived elliptic motion via the Law of Gravity.) Except for passing references in subsequent chapters, abduction via new concepts will be outside the scope of our analysis.

Abductive Processes

Once the above parameters get set, several kinds of abductive processes arise. For example, abduction triggered by novelty with an underlying deductive inference, calls for a process by which the theory is expanded with an abductive explanation. The fact to be explained is consistent with the theory, so an abductive explanation added to the theory accounts deductively for the fact. However, when the underlying inference is statistical, in a case of novelty, theory expansion might not be enough. The added statement might lead to a 'marginally consistent' theory with low probability, which would not yield a strong explanation for the observed fact. In such a case, theory revision is needed (i.e. removing some data from the theory) to account for the observed fact with high probability. (For a specific example of this latter case cf. chapter 8.)

Our aim is not to classify abductive processes, but rather to point out that several kinds of these are used for different combinations of the above parameters. In the coming chapters we explore in detail some procedures for computing different types of outcomes in a deductive format; those triggered by novelty (chapter 4) and those triggered by anomaly (chapter 8).

Examples Revisited

Given our taxonomy for abductive reasoning, we can now see more patterns across our earlier list of examples. Varying the inferential parameter, we cover not only cases of deduction but also statistical inferences. Thus, Hempel's statistical model of explanation [Hem65] becomes a case of an (abductive) explanatory argument. Our example (4) of medical diagnosis (Jane's quick recovery) was an instance. Logic programming inference seems more appropriate to an example like (1) (whose overall structure is similar to (4)). As for triggers, novelty drives both rain examples (1) and (2), as well as the medical diagnosis one (4). A trigger by anomaly occurs in example (3), where the theory predicts the contrary of our observation, the lights-off example (5), and the Kepler example (6), since his initial observation of the longitudes of Mars contradicted the previous rule of circular orbits of the planets. As for different outcomes, examples (1), (3), (4) and (5) abduce facts, examples (2) and (6) produce rules as their forms of explanantia. Different forms of outcomes will play a role in different types of procedures for producing abductive explanations. In computer science jargon, triggers and outcomes are, respectively, preconditions and outputs of abductive devices, whether these be computational procedures or inferential ones.

This taxonomy gives us the big picture of abductive reasoning. In the remainder of this book, we are going to investigate several of its aspects, which give rise to more specific logical and computational questions. (Indeed, more than we have been able to answer!) Before embarking upon this course, however, we need to discuss one more general strategic issue, which explains the separation of concerns between chapters 3 and 4 that are to follow.

Abductive Logic: Inference + Search Strategy

Classical logical systems have two components: a semantics and a proof theory. The former aims at characterizing what it is for a formula to be true in a model, and it is based on the notions of truth and interpretation. The latter characterizes what counts as a valid proof for a formula, by providing the inference rules of the system; having for its main notions proof and derivability. These two formats can be studied independently, but they are closely connected. At least in classical (first-order) logic, the completeness theorem tells us that all valid formulas have a proof, and vice versa. Many logical systems have been proposed that follow this pattern: propositional logic, predicate logic, modal logic, and various typed logics.

From a modern perspective, however, there is much more to reasoning than this. Computer science has posed a new challenge to logic; that of providing automatic procedures to operate logical systems. This requires a further fine-structure of reasoning. In fact, recent studies in AI give precedence to a *control strategy* for a logic over its complete proof theory. In particular, the heart of logic programming lies in its control strategies, which lead to much greater sensitivity as to the order in which premises are given, the avoidance of search loops, or the possibility to cut proof searches (using the extra-logical operator !) when a solution has been found. These features are extralogical from a classical perspective, but they do have a clear formal structure, which can be brought out, and has independent interest as a formal model for broader patterns of argumentation (cf. [vBe92, Kal95, Kow91]).

Several contemporary authors stress the importance of control strategies, and a more finely-structured algorithmic description of logics. This concern is found both in the logical tradition ([Gab94a, vBe90]), and in the philosophical tradition ([Gil96]), the latter arguing for a conception of logic as: *inference + control*. (Note the shift here away from Kowalski's famous dictum "Algorithm = Logic + Control".) In line with this philosophy, we wish to approach abduction with two things in mind. First, there is the inference parameter, already discussed, which may have several interpretations. But given any specific choice, there is still a significant issue of a suitable search strategy over this inference, which models some particular abductive practice. The former parameter may be defined in semantic or proof-theoretic terms. The search procedure, on the other hand, deals with concrete mechanisms for producing valid inferences. It is then possible to control which kinds of outcome are produced with a certain efficiency. In particular, in abduction, we may want to produce only 'useful' or 'interesting' formulas, preferably even just some 'minimal set' of these.

In this light, the aim of an abductive search procedure is not necessarily completeness with respect to some semantics. A procedure that generates all possible abductive explanations might be of no practical use, and might also miss important features of human abductive activity. In chapter 4, we are going to experiment with semantic tableaux as a vehicle for attractive abductive strategies that can be controlled in various ways.

PART II

LOGICAL FOUNDATIONS

Chapter 3

ABDUCTION AS LOGICAL INFERENCE

1. Introduction

In the preceding overview chapter, we have seen how the notion of abduction arose in the last century out of philosophical reflection on the nature of human reasoning, as it interacts with patterns of explanation and discovery. Our analysis brought out a number of salient aspects to the abductive process, which we shall elaborate in a number of successive chapters. For a start, abduction may be viewed as a kind of logical inference and that is how we will approach it in the analysis to follow here. Evidently, though, as we have already pointed out, it is not standard logical inference, and that for a number of reasons. Intuitively, abduction runs in a *backward* direction, rather than the *forward* one of standard inference, and moreover, being subject to revision, it exhibits non-standard non-monotonic features (abductive conclusions may have to be retracted in the light of further evidence), that are more familiar from the literature on non-standard forms of reasoning in artificial intelligence. Therefore, we will discuss abduction as a broader notion of consequence in the latter sense, using some general methods that have been developed already for non-monotonic and dynamic logics, such as systematic classification in terms of structural rules. This is not a mere technical convenience. Describing abduction in an abstract general way makes it comparable to better-studied styles of inference, thereby increasing our understanding of its contribution to the wider area of what may be called 'natural reasoning'. To be sure, in this chapter we propose a logical characterization of what we have called an *(abductive) explanatory argument*, in order to make explicit that the inference is explanatory (and thus forward chained), while keeping in mind it aims to characterize the conditions for an abductive explanation to be part of this inference (cf. chapter 2).

The outcomes that we obtain in this first systematic chapter, naturally divided into five parts, are as follows. After this introduction, in the second part (section 2), we discuss the problem of demarcation in logic, in order to set the ground for our analysis of abduction as a logical inference. Placing abduction in a broader universe of logics and stances, make natural to consider it as a logical inference of its own kind. In the third part (section 3), we discuss in detail some aspects of abductive inference, such as its direction, format of premisses and conclusion as well as its inferential strength. We propose a general logical format for abduction, involving more parameters than in standard inference, allowing for genuinely different roles of premisses. We find a number of natural styles of abductive explanatory inference, rather than one single candidate. These (abductive) explanatory versions are classified by different *structural rules* of inference, and this issue occupies the fourth part (section 4). As a contribution to the logical literature in the field, we give a complete characterization of one simple style of abductive explanatory inference, which may also be viewed as the first structural characterization of a natural style of explanation in the philosophy of science. We then analyze some other abductive explanatory versions (explanatory, minimal and preferential) with respect to their structural behaviour, giving place to more sophisticated structural rules with interest of their own. Finally, we turn to discuss further logical issues such as how those representations are related to more familiar completeness theorems, and finally, we show how abduction tends to involve higher complexity than classical logic: we stand to gain more explanatory power than what is provided by standard inference, but this surplus comes at a price. In the fifth and final part of this chapter (section 5),we offer an analysis of previous sections centering the discussion on abduction as an enriched form of logical inference with an structure of its own. We then put forward our conclusions and present related work within the study of non-monotonic reasoning.

Despite these useful insights, pure logical analysis does not exhaust all there is to abduction. In particular, it's more dynamic process aspects, and its interaction with broader conceptual change must be left for subsequent chapters, that will deal with computational aspects, as well as further connections with the philosophy of science and artificial intelligence. What we do claim, however, is that our logical analysis provides a systematic framework for studying the intuitive notion of abductive explanatory inference, which gives us a view of its variety and complexity, and which allows us to raise some interesting new questions.

2. Logic: The Problem of Demarcation

One of the main questions for logic is the problem of demarcation, that is, to distinguish between logical and non-logical systems. This question is at the core of the philosophy of logic, and has a central place in the philosophy of mathematics, in the philosophy of science as well as in the foundations of artificial intelligence.

Some questions in need for an answer for this problem concern the following ones: what is a logic?, which is the scope of logic?, which formal systems qualify as logics?, all of these leading to metaphysical questions concerning the notion of correctness of a logical system: does it make sense to speak of a logical system as correct or incorrect?, could there be several logical systems which are equally correct?, is there just one correct logical system? These questions in turn lead to epistemological questions of the following kind: how does one recognise a truth of logic? could one be mistaken in what one takes to be such truths?

There are however, several proposals and positions in the literature in regard to all these questions. Our strategy to describe the problem of demarcation of logic will be the following. Our point of departure is Peirce's distinction of three types of reasoning, namely deduction, induction, and abduction. We will compare them according to their certainty level, something that in turn gives place to different areas of application, mainly in mathematics, philosophy of science and artificial intelligence. Next, we will introduce an standard approach in philosophy of logic based on the relationship between informal arguments and their counterparts in formal logic, namely the view endorsed by Haack [Haa78]. Her classification of kinds of logics will be presented, that is, the well-known distinction amongst extensions and deviations of classical logics, and inductive logics. Moreover, we take up on Haack's discussion on the several positions with respect to the legitimization (and proliferation) of logics, namely instrumentalism, monism and pluralism. Finally, we will introduce a much less-known approach –but still standard within its field – coming from artificial intelligence, namely the logical structural approach devised for the study of non-monotonic reasoning.

Our overall discussion in this section will serve two purposes. On the one hand, it aims to show that even under a broad view of logic, there is neither a unique nor a definite answer to the problem of demarcation, not to mention to each of the former questions. On the other hand, it will set the ground for the main purpose of this chapter, that is, an analysis of abduction as a specific kind of logical inference, in order to show, that abduction holds a natural place to be considered a logical inference of its own kind.

Types of Reasoning: Deduction, Induction, Abduction

From a logical perspective, mathematical reasoning may be identified with classical, deductive inference. Two aspects are characteristic of this type of reasoning, namely its *certainty* and its *monotonicity*. The first of these is exemplified by the fact that the relationship between premisses and conclusion is that of necessity; a conclusion drawn from a set of premisses, necessarily follows from them. The second aspect states that conclusions reached via deductive reasoning are non-defeasible. That is, once a theorem has been proved, there is no doubt of its validity regardless of further addition of axioms and theorems to the system.

There are however, several other types of formal non-classical reasoning, which albeit their lack of complete certainty and monotonicity, are nevertheless rigorous forms of reasoning with logical properties of their own. Such is the case of inductive and abductive reasoning. As a first approximation, Charles S. Peirce distinction seems useful. According to him, there are three basic types of logical reasoning: *deduction*, *induction* and *abduction*. Concerning their certainty level: *'Deduction proves that something must be; Induction shows that something actually is operative; Abduction merely suggests that something may be'* [CP, 5.171]. Therefore, while deductive reasoning is completely certain, inductive and abductive reasoning are not. *'Deduction is the only necessary reasoning. It is the reasoning of mathematics'* [CP, 4.145]. Induction must be validated empirically with tests and experiments, therefore it is defeasible; and abductive reasoning can only offer hypotheses that may be refuted with additional information. For example, a generalization reached by induction (e.g. all birds fly), remains no longer valid after the addition of a premisse, which refutes the conclusion (e.g. penguins are birds). As for abduction, a hypothesis (e.g. it rained last night) which explains an observation (e.g. the lawn is wet), may be refuted when additional information is incorporated into our knowledge base (e.g. it is a drought period).

Deductive reasoning has been the paradigm of mathematical reasoning, and its logic is clearly identified with Tarski's notion of logical inference. In contrast, inductive and abductive types of reasoning are paradigmatic types of reasoning in areas like philosophy of science, and more recently, artificial intelligence. Regarding the former, contemporary research indicates that many questions regarding their logic remain controversial. As it is well known, Carnap's proposal for an inductive logic[Car55] found ample criticisms. As for abduction, while some scholars argue that the process of forming an explanatory hypothesis (our abductive process) cannot be logically reconstructed [Pop59, Hem65], and have instead proposed each a logical characterization of explanation (our (abductive) explanatory argument)[1]; others have tried to formally characterize

[1] As for the roots and similarities of these two models of explanation, Niiniluoto[Nii00, p. 140] rightly observes: *"After Hempel's (1942) paper about the deductive–nomological pattern of historical explanation, Karl Popper complained that Hempel had only reproduce his theory of causal explanation, originally presented in 'Logik der Forschung' (1935, see Popper 1945, chap 25, n. 7; Popper 1957, p. 144). With his charming politeness, Hempel pointed out that his account of D–N explanation is 'by no means novel' but 'merely summarizes and states explicitly some fundamental points which have been recognized by many scientists and methodologists"*.

'retroduction' (another term for abduction), as a form of inversed deduction [Han61], but no acceptable formulation has been found. Regarding the latter, recent logico-computational oriented research has focus on studying non-standard forms of reasoning, in order to build computer programs modeling human reasoning, which being subject to revision, is uncertain and exhibits non-standard non-monotonic features. Several contemporary authors propose a more finely structured algorithmic description of logics. This concern is found both in the logical tradition ([Gab94a, vBe90]), as well as in work in philosophy ([Gil96]).

Logics: Extensions, Deviations, Inductive

Haack[Haa78] takes as primitive an intuitive notion of a formal system, and from there it hints at the characterization of what is to be a logical system, as follows:

> *"The claim for a formal system to be a logic depends, I think, upon its having an interpretation according to which it can be seen as aspiring to embody canons of valid argument."* [Haa78, page 3].

The next problem to face is that of deciding what counts as valid argumentation. But before we get into her own answer to this question, here are other criteria aiming to characterize what counts as a logical system. On the one hand, according to Kneale, logical systems are those that are purely formal, for him, those that are *complete* (in which all universal valid formula are theorems). According to Dummet, on the other hand, logical systems are those which characterize *precise* notions. Following the first characterization, many formal systems are left out, such as second order logic. If we follow the second one, then proposals such as Hintikka's system of epistemic logic is left out as well, for the notions of knowledge and belief characterize pretty vague epistemic concepts [Haa78, page 7]. Both these characterizations provide purely formal criteria for logical demarcation. For Haack, however *'the prospects for a well-motivated formal criterion are not very promising'*[Haa78, page 7], for it has the drawback of limiting the scope of logic to the point of even discarding well accepted formal systems (e.g. predicate logic) on the basis of being in absence of other metalogical properties (e.g. decidability). Moreover, many logical systems are indeed undecidable, incomplete, but nevertheless have interesting applications and have proved useful in areas like computer science and linguistics.

Haack takes a broad view of logic, considering that *'the demarcation is not based on any very profound ideas about 'the essential nature of logic'*[Haa78, page 4], and follows 'the benefit of the doubt policy', according to which, arguments may be assessed by different standards of validity, and thus accepts several formal systems as logical. For her, the question we should be asking is whether a system is good and useful rather than 'logical', which after all is not a well-defined concept. Her approach however, is not wholly arbitrary, for it does not give up the requirement of being rigorous, and takes classical logic as its reference point, building up a classification of systems of logic based on analogies to the classical system, as follows:

- Extensions (*e*)

  Modal, Epistemic, Eroretic, ...

- Deviations (*d*)

  Intuitionistic, Quantum, Many-valued, ...

- Inductive (*i*)

  Inductive probability logic

Extensions (*e*) are formal logical systems, which extend the system of classical logic ($\mathcal{L}_c$) in three respects: their language, axioms and rules of inference ($\mathcal{L}_c \subseteq \mathcal{L}_e, \mathcal{A}_c \subseteq \mathcal{A}_e, \mathcal{R}_c \subseteq \mathcal{R}_e$). These systems preserve all valid formula of the classical system, and therefore all previous valid formula remain valid as well ($\forall\varphi(\Sigma \models_c \varphi \Rightarrow \Sigma \models_e \varphi); \varphi \in \mathcal{L}_c$). So, for instance, modal logic extends classical system by the modal operators of necessity and possibility together with axioms and rules for them.

Deviations (*d*) are formal systems that share the language with the system of classical logic ($\mathcal{L}_c$) but that deviate in axioms and rules ($\mathcal{L}_c = \mathcal{L}_d, \mathcal{A}_c \neq \mathcal{A}_d, \mathcal{R}_c \neq \mathcal{R}_d$). Therefore, some formulae, which are valid in the classical system, are no longer valid in the deviant one ($\exists\varphi(\Sigma \models_c \varphi \wedge \Sigma \not\models_d \varphi); \varphi \in \mathcal{L}_c$). Such is the case of intuitionistic logic, in which the classical axiom $A \vee \neg A$ is no longer valid.

Inductive systems (*i*) are formal systems that share the language with the system of classical logic ($\mathcal{L}_c = \mathcal{L}_i$), but in which no formula which is valid by means of the inductive system is valid in the classical one ($\forall\varphi(\Sigma \models_i \varphi \Rightarrow \Sigma \not\models_c \varphi); \varphi \in \mathcal{L}_i$). Here the basis is the notion of 'inductive strength', and the idea is that 'an argument is inductively strong if its premisses give a certain degree of support, even if less than conclusive support, to its conclusion: if, that is, it is *improbable* that its premisses (Σ) should be true and its conclusion (φ) false' ($\text{NOT PROB}(\Sigma \wedge \neg\varphi)$) [Haa78, page 17].

In each of these logical systems there is an underlying notion of logical consequence (or of derivability), which settles the validity of an argument within the system. While the first two categories pertain to formal systems which are deductive in nature, the third one concerns inductive ones. But still there may be several characterizations for both deductive and inductive kinds. For example, one deviant system, that of relevance logic renders the notion of classical consequence insufficient and asks for more: an argument in relevance logic must meet the requirement that the premisses be 'relevant' to its conclusion. As for inductive systems, another way of characterizing them is that for which 'it is improbable, *given that* the premisses (Σ) are true, that the conclusion is false ($\neg\varphi$)' [Haa78, page 17]. We may interpret this statement in terms of conditional probability as follows[2]: ($\text{NOT PROB}(\Sigma/\neg\varphi)$). Notice that deductive validity is a limiting case of inductive strength, where the probability of the premisses being true and the conclusion false is cero, for the first characterization, and where it is certain that the conclusion is true when the premisses are, for this second one.

In the overall, under this approach arguments may be assessed by deductive or inductive standards, and thus there may be deductively valid, inductively strong or neither. As we shall see later in this chapter (cf. section 5.1), this classification does not explicitly take into account abductive logic, but it can nevertheless be accommodated within.

Positions: Instrumentalism, Monism, Pluralism

The position taken with respect to the demarcation of logic largely depends upon the answers given to metaphysical questions concerning the notion of correctness of a logical system, which in turn depend on the distinction between system-relative and extra-systematic validity/logical truth. Roughly speaking, a logical system is correct if the formal arguments (and formula) which are valid (logically true) in that system correspond to informal arguments (statements), which are valid (logically true) in the extra-systematic sense([Haa78, page 222]). Three positions are characterized by Haack, each of which is characterized by the answers (affirmative or negative) given to the following questions:

Questions:

- A: Does it make sense to speak of a logical system as correct or incorrect?
- B: Are there extra-systematic conceptions of validity/logical truth by means of which to characterize what is it for a logic to be correct?
- C: Is there one correct system?
- D: Must a logical system aspire to global application, i.e. to represent reasoning irrespective of subject-matter, or may a logic be locally correct. i.e. correct within a limited area of discourse?

---

[2]I thank the anonymous referee for this particular suggestion.

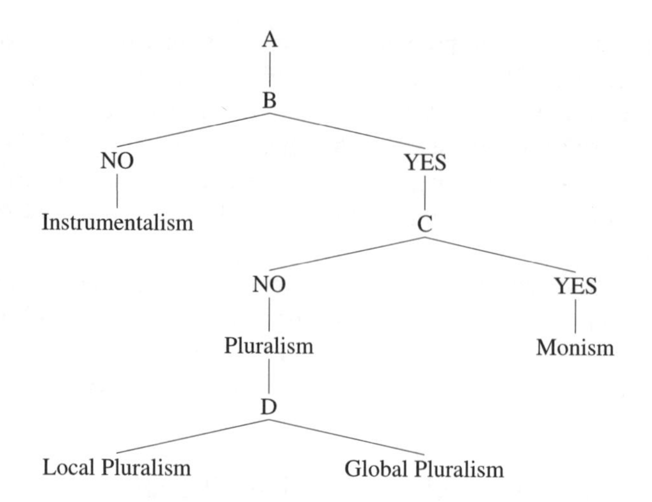

Thus, on the one hand, the instrumentalist position answers the first two questions negatively. It is based on the idea that the notion of 'correctness' for a system is inappropriate, and that one should rather be asking for its being more fruitful, useful, convenient... etc than another one. 'An instrumentalist will only allow the 'internal' question, whether a logical system is *sound*, whether, that is, all and only the theorems/syntactically valid arguments *of the system* are logically true/valid *in the system*'[Haa78, page 224]. On the other hand, both the monist and the pluralist answer these questions in the affirmative, the difference being that while the monist recognizes one and only one system of logic, the pluralist accepts a variety of them. Thus, they answer the third question opposite. Note that the distinction in these questions is only relevant for the classical logic vs. deviant logic dichotomy. The reason being that for a monist classical logic and its extensions are fragments of a 'correct system', and for a pluralist classical logic and its extensions are both 'correct'.

Likewise, while for an instrumentalist there are not extra-systematic conceptions of validity/logical truth by means of which to characterize what is to be a logic to be correct, for the monist as well as for the pluralist there are, either in the unitary fashion or in the pluralistic one. A further distinction made by the pluralist concerns the scope of application for a certain logical system. While a global pluralist endorses the view that a logical system must aspire to represent reasoning irrespective of subject-matter, a local pluralist supports the view that a logical system is only locally correct within a limited area of discourse.

The next question to analyze is the position taken by each of these stances with regard to whether deviant logics rival classical logic. In order to answer this question we have the following diagram:

(i) Formal argument which is	(ii) Valid in L
represents	corresponding to (iii)'s being
(iii) Informal argument	(iv) extra-systematically valid

On the one hand, the monist answers this question in the affirmative and supports the view that (i) aspires to represent (iii) in such a way that (ii) and (iv) do correspond in the 'correct logic'. On the other hand, the local pluralist answers this question in the negative by relativizing (iv) to specific areas of discourse and the global pluralist either fragments the relation between (i) and (iii) (that is, denies that the formal arguments of a deviant system represent the same informal arguments as those of classical logic) or fragments the relationship between (ii) and (iv) (denies that validity in the deviant logic is intended to correspond to extra-systematic validity as that to which validity in classical logic is intended to correspond). Finally, the instrumentalist rejects (iv) altogether.

Structural Logical Approach

This type of analysis (started in [Sco71]) was inspired in the works of logical consequence by Tarski [Tar83] and those of natural deduction by Gentzen [SD93, Pao02]. It describes a style of inference at a very abstract structural level, giving its pure combinatorics. It has proved very successful in artificial intelligence for studying different types of plausible reasoning ([KLM90]), and indeed as a general framework for non-monotonic consequence relations ([Gab85]). Another area where it has proved itself is dynamic semantics, where not one but many new notions of dynamic consequences are to be analyzed ([vBe96a]). The basic idea of logical structural analysis is the following:

> A notion of logical inference can be completely characterized by its basic combinatorial properties, expressed by structural rules.

Structural rules are instructions which tell us, e.g., that a valid inference remains valid when we insert additional premisses ('monotonicity'), or that we may safely chain valid inferences ('transitivity' or 'cut'). To understand this perspective in more detail, one must understand how it characterizes classical inference. In what follows we use logical sequents with a finite sequence of premisses to the left, and one conclusion to the right of the sequent arrow ($\Sigma \Rightarrow C$). While X, Y and Z are finite sets of formulae, A, B and C are single formula.

Classical Inference

The structural rules for classical inference are the following:

- **Reflexivity:** $C \Rightarrow C$
- **Contraction:**

$$\frac{X, A, Y, A, Z \Rightarrow C}{X, A, Y, Z \Rightarrow C}$$

- **Permutation:**

$$\frac{X, A, B, Y \Rightarrow C}{X, B, A, Y \Rightarrow C}$$

- **Monotonicity:**

$$\frac{X, Y \Rightarrow C}{X, A, Y \Rightarrow C}$$

- **Cut Rule:**

$$\frac{X, A, Y \Rightarrow C \quad Z \Rightarrow A}{X, Z, Y \Rightarrow C}$$

These rules state the following properties of classical consequence. Any premisse implies itself (reflexivity), deleting repeated premisses causes no trouble (contraction); premisses may be permuted without altering validity (permutation), adding new information does not invalidate previous conclusions (monotonicity), and premisses may be replaced by sequences of premisses implying them (cut). In all, these rules allow us to treat the premisses as a mere set of data without further relevant structure. This plays an important role in classical logic, witness what introductory textbooks have to say about "simple properties of the notion of consequence"[3]. Structural rules are also used extensively in completeness proofs[4].

These rules are structural in that they mention no specific symbols of the logical language. In particular, no connectives or quantifiers are involved. This makes the structural rules different from inference rules like, say, Conjunction of Consequents or Disjunction of Antecedents, which also fix the meaning of

[3] In [Men64, Page 30] the following simple properties of classical logic are introduced:

- If $\Gamma \subseteq \Delta$ and $\Gamma \vdash \phi$, then $\Delta \vdash \phi$.
- $\Gamma \vdash \phi$ iff there is a finite subset Δ of Γ such that $\Delta \vdash \phi$.
- If $\Gamma \vdash x_i$ (for all i) and $x_1, \ldots, x_n \vdash \phi$ then $\Gamma \vdash \phi$.

Notice that the first is a form of Monotonicity, and the third one of Cut.

[4] As noted in [Gro95, page46]: "In the Henkin construction for first-order logic, or propositional modal logic, the notion of maximal consistent set plays a major part, but it needs the classical structural rules. For example, Permutation, Contraction and Expansion enable you to think of the premisses of an argument as a set; Reflexivity is needed to show that for maximal consistent sets, membership and derivability coincide".

conjunction and disjunction. Under this approach, Haack's previous classification of extensions of logics is subsumed, for one rule may fit classical logic as well as extensions: propositional, first-order, modal, type-theoretic, etc. Each rule in the above list reflects a property of the set-theoretic definition of classical consequence ([Gro95]), which – with some abuse of notation – calls for inclusion of the intersection of the (models for the) premisses in the (models for the) conclusion:

$P_1, \ldots, P_n \Rightarrow C$ iff $P_1 \cap \ldots \cap P_n \subseteq C$.

Now, in order to prove that a set of structural rules *completely* characterizes a style of reasoning, representation theorems exist. For classical logic, one version was proved by van Benthem in [vBe91]:

PROPOSITION 1 *Monotonicity, Contraction, Reflexivity, and Cut completely determine the structural properties of classical consequence.*

Proof. Let R be any abstract relation between finite sequences of objects and single objects satisfying the classical structural rules. Now, define:

$$a^* = \{A \mid A \text{ is a finite sequence of objects such that ARa}\}.$$

Then, it is easy to show the following two assertions:

1 If $a_1, \ldots, a_k Rb$, then $a_1^* \cap \ldots \cap a_k^* \subseteq b^*$, using Cut and Contraction.

2 If $a_1^* \cap \ldots \cap a_k^* \subseteq b^*$, then $a_1, \ldots, a_k Rb$, using Reflexivity and Monotonicity. ⊣

Permutation is omitted in this theorem. And indeed, it turns out to be derivable from Monotonicity and Contraction.

We have thus shown that classical deductive inference, observes easy forms of reflexivity, contraction, permutation, monotonicity and cut. The representation theorem shows that these rules completely characterize this type of reasoning.

Non-Classical Inference

For non-classical consequences, classical structural rules may fail. Well-known examples are the ubiquitous 'non-monotonic logics'. However, this is not to say that no structural rules hold for them. The point is rather to find appropriate reformulations of classical principles (or even entirely new structural rules) that fit other styles of consequence. For example, many non-monotonic types of inference do satisfy a weaker form of monotonicity. Additions to the premisses are allowed only when these premisses imply them:

- Cautious Monotonicity:

$$\frac{X \Rightarrow C \qquad X \Rightarrow A}{X, A \Rightarrow C}$$

Dynamic inference is non-monotonic (inserting arbitrary new processes into a premisse sequence can disrupt earlier effects). But it also quarrels with other classical structural rules, such as Cut. But again, representation theorems exist. Thus, the 'update-to-test' dynamic style (once in which a process cannot be disrupted) and is characterized by the following restricted forms of monotonicity and cut, in which additions and omissions are licensed only to the left side:

- Left Monotonicity:

$$\frac{X \Rightarrow C}{A, X \Rightarrow C}$$

- Left Cut:

$$\frac{X \Rightarrow C \qquad X, C, Y \Rightarrow D}{X, Y \Rightarrow D}$$

For a broader survey and analysis of dynamic styles, see [Gro95, vBe96a]. For sophisticated representation theorems in the broader field of non-classical inference in artificial intelligence see [Mak93, KLM90]. Yet other uses of non-classical structural rules occur in relevance logic, linear logic, and categorial logics (cf. [DH93, vBe91]. [Gab94b]).

Characterizing a notion of inference in this way, determines its basic repertoire for handling arguments. Although this does not provide a more ambitious semantics, or even a full proof theory, it can at least provide valuable hints. The suggestive Gentzen style format of the structural rules turns into a sequent calculus, if appropriately extended with introduction rules for connectives. However, it is not always clear how to do so in a natural manner, as we will discuss later in connection with abduction.

The structural analysis of a logical inference is a metalevel explication which is based on structural rules and not on language, as it does not take into account logical connectives or constants, and in this respect differs from Haack's approach.

3. Abductive Explanatory Argument: A Logical Inference

Here are some preliminary remarks about the logical nature of abductive inference, which set the scene for our subsequent discussion. The standard textbook pattern of logical inference is this: *Conclusion C follows from a set of premisses P*. This format has its roots in the axiomatic tradition in mathematics that follows the deductive method, inherited from Euclid's *Elements*, in which from a certain set of basic axioms, all geometrical truths of elementary geometry are derived. This work is not only the first logical system of its kind, but it has been the model to follow in mathematics as well as in other formal scientific enterprises. Each proposition is linked, via proofs, to previous axioms, definitions and propositions. This method is forward chained, picturing a reasoning from a finite set of premisses to a conclusion, and it is completely certain and monotonic[5].

Moreover, there are at least two ways of thinking about validity in this setting, one semantic, based on the notions of model and interpretation (every model in which P is true makes C true), the other syntactic, based on a proof-theoretic derivation of C from P. Both explications suggest forward chaining from premisses to conclusions: $P \Rightarrow C$ and the conclusions generated are undefeasible. We briefly recall some features that make abduction a form of inference that does not fit easily into this format. All of them emerged in the course of our preceding chapter. Most prominently, in abduction, the conclusion is the given and the premisses (or part of them) are the output of the inferential process: $P \Leftarrow C$. Moreover, the *abduced* premisse has to be *consistent* with the background theory of the inference, as it has to be *explanatory*. And such explanations may undergo change as we modify our background theory. Finally, when different sets of premisses can be abduced as explanations, we need a notion of preference between them, allowing us to choose a *best* or *minimal* one. These various features, though non-standard when compared with classical logic, are familiar from neighbouring areas. For instance, there are links with classical accounts of explanation in the philosophy of science [Car55, Hem65], as well as recent research in artificial intelligence on various notions of common sense reasoning [McC80, Sho88, Gab96]. It has been claimed that this is an appropriate broader setting for general logic as well [vBe90], gaping back to the original program by Bernard Bolzano (1781–1848), in his "Wissenschaftslehre" [Bol73]. Indeed, our discussion of abduction in Peirce in the preceding chapter reflected a typical feature of pre-Fregean logic: boundaries between logic and general methodology were still rather fluid. In our view, current post-Fregean logical research is slowly moving back towards this same broader agenda. More concretely, we shall review the mentioned features of abduction in some further detail now, making a few strategic references to this broader literature.

Directions in Reasoning: Forward and Backward

Intuitively, a valid inference from, say, premisses P_1, P_2 to a conclusion C allows for various directions of thought. In a forward direction, given the premisses, we may want to draw some strongest, or rather, some most appropriate conclusion. (Notice incidentally, that the latter notion already introduces a certain dependence on context, and good sense: the strongest conclusion is simply

[5]This is not say however –as Hilbert would have liked to claim– that all mathematical reasoning may be reduced to axiomatics. As it is well know by the incompleteness results of Gödel, there are clear limitations to reasoning in mathematics through the axiomatic method. Moreover, the view of mathematics as an experimental, empirical science, found in philosophy[Lak76] as well as in recent work in computer science [Cha97], shows that axiomatics cannot exhaust all there is to mathematical reasoning.

$P_1 \wedge P_2$, but this will often be unsuited.) Classical logic also has a backward direction of thought, when engaged in refutation. If we know that C is false, then at least one of the premisses must be false. And if we know more, say the truth of P_1 and the falsity of the conclusion, we may even refute the specific premisse P_2. Thus, in classical logic, there is a duality between forward proof and backward refutation. This duality has been noted by many authors. It has even been exploited systematically by Beth when developing his refutation method of semantic tableaux [Bet69]. Read in one direction, closed tableaux are eventually failed analyses of possible counterexamples to an inference, read in another they can be arranged to generate a Gentzen-style sequent derivations of the inference (we shall be using tableaux in our next chapter, on computing abduction.) Beth's tableaux can be taken as a formal model of the historical opposition between methods of 'analysis' and 'synthesis' in the development of scientific argument (cf. chapter 1). Methodologically, the directions are different sides of the same coin, namely, some appropriate notion of inference.

Likewise, in abduction, we see an interplay of different directions. This time, though, the backward direction is not meant to provide refutations, but rather confirmations. We are looking for suitable premisses that would support the conclusion[6].

Our view of the matter is the following. In the final analysis, the distinction between directions is a relative one. What matters is not the direction of abduction, but rather an interplay of two things. As we have argued in chapter 2, one should distinguish between the choice of an underlying *notion of inference* $\Rightarrow$, and the independent issue as to the *search strategy* that we use over this. Forward reasoning is a bottom up use of $\Rightarrow$, while backward reasoning is a top-down use of $\Rightarrow$. In line with this, in this chapter, we shall concentrate on notions of inference $\Rightarrow$ leaving further search procedures to the next, more computational chapter 4. In this chapter the intuitively backward direction of abduction is not crucial to us, except as a pleasant manner of speaking. Instead, we concentrate on appropriate underlying notions of consequence for abduction.

Formats of Inference: Premisses and Background Theory

The standard format of logical inference is essentially binary, giving a transition from premisses to a conclusion:

$$\frac{P_1, \ldots, P_n}{C}$$

[6]In this case, a corresponding refutation would rather be a forward process: if the abduced premisse turns out false, it is discarded and an alternative hypothesis must be proposed. Interestingly, [Tij97] (a recent practical account of abduction in diagnostic reasoning) mixes both 'positive' confirmation of the observation to be explained with 'refutation' of alternatives.

These are 'local steps', which take place in the context of some, implicit or explicit, background theory (as we have seen in chapter 2). In this standard format, the background theory is either omitted, or lumped together with the other premisses. Often this is quite appropriate, especially when the background theory is understood. But sometimes, we do want to distinguish between different roles for different types of premisse, and then a richer format becomes appropriate. The latter have been proposed, not so much in classical logic, but in the philosophy of science, artificial intelligence, and informal studies on argumentation theory. These often make a distinction between explicit premisses and implicit *background assumptions*. More drastically, premisse sets, and even background theories themselves often have a hierarchical structure, which results in different 'access' for propositions in inference. This is a realistic picture, witness the work of cognitive psychologists like [Joh83].

In Hempel's account of scientific explanation (cf. chapter 5) premisses play the role of either scientific laws, or of initial conditions, or of specific explanatory items, suggesting the following format:

$$\begin{array}{c}\text{Scientific laws + initial conditions + explanatory facts}\\ \Downarrow \\ \text{Observation}\end{array}$$

Further examples are found on the borderline of the philosophy of science and philosophical logic, in the study of conditionals. The famous 'Ramsey Test' presupposes revision of explicit beliefs in the background assumptions [Sos75, vBe94], which again have to be suitably structured. More elaborate hierarchical views of theories have been proposed in artificial intelligence and computer science. [Rya92] defines 'ordered theory presentations', which can be arbitrary rankings of principles involved in some reasoning practice. (Other implementations of similar ideas use labels for formulas, as in the labelled deductive systems of [Gab96].) While in Hempel's account, structuring the premisses makes sure that scientific explanation involves an interplay of laws and facts, Ryan's motivation is resolution of conflicts between premisses in reasoning, where some sentences are more resistant than others to revision. (This motivation is close to that of the Gärdenfors theory, to be discussed in chapter 8. A working version of these ideas is found in a study of abduction in diagnosis ([Tij97], which can be viewed as a version of our later account in this chapter with some preference structure added.) More structured views of premisses and theories can also be found in situation semantics, with its different types of 'constraints' that govern inference (cf. [PB83]).

In all these proposals, the theory over which inference takes place is not just a bag into which formulas are thrown indiscriminately, but an organized structure in which premisses have a place in a hierarchy, and play specific different roles. These additional features need to be captured in richer inferential formats for more complicated reasoning tasks. Intuitive 'validity' may be partly based on the type and status of the premisses that occur. We cite one more example, to elaborate what we have in mind.

In argumentation theory, an interesting proposal was made in [Tou58]. Toulmin's general notion of consequence was inspired on the kind of reasoning done by lawyers, whose claims need to be defended according to juridical procedures, which are richer than pure mathematical proof. Toulmin's format of reasoning contains the necessary tags for these procedures:

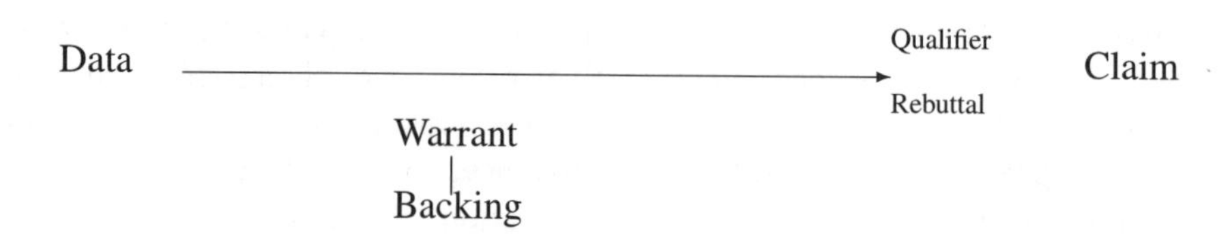

Every claim is defended from certain relevant data, by citing (if pressed) the background assumptions (one's 'warrant') that support this transition. (There is a dynamic process here. If the warrant itself is questioned, then one has to produce one's further 'backing'.) Moreover, indicating the purported strength of the inference is part of making any claim (whence the 'qualifier'), with a 'rebuttal' listing some main types of possible exception (rebuttal) that would invalidate the claim. [vBe94] relates this format to issues in artificial intelligence, as it seems to describe common sense reasoning rather well. Toulmin's model has also been proposed as a mechanism for intelligent systems performing explanation ([Ant89]).

Thus, once again, to model reasoning outside of mathematics, a richer format is needed. Notice that the above proposals are syntactic. It may be much harder to find purely semantic correlates to some of the above distinctions: as they seem to involve a reasoning procedure rather than propositional content. (For instance, even the distinction between individual facts and universal laws is not as straightforward as it might seem.) Various aspects of the Toulmin schema will return in what follows. For Toulmin, inferential strength is a parameter, to be set in accordance with the subject matter under discussion. (Interestingly, content-dependence of reasoning is also a recurrent finding of cognitive psychologists: cf. the earlier-mentioned [Joh83].) In chapter 2, we have already defended exactly the same strategy for abduction. Moreover, the procedural flavor of the Toulmin schema fits well with our product-process distinction.

As for the basic building blocks of abductive explanatory inference, in the remainder of this book, we will confine ourselves to a ternary format:

$$\Theta \mid \alpha \Rightarrow \varphi$$

This modest step already enables us to demonstrate a number of interesting departures from standard logical systems. Let us recall some considerations from chapter 2 motivating this move. The theory Θ needs to be explicit for a number of reasons. Validity of an abductive inference is closely related to the background theory, as the presence of some other explanation β in Θ may actually disqualify α as an explanation. Moreover, what we called 'triggers' of explanation are specific conditions on a theory Θ and an observation φ. A fact may need explanation with respect to one theory, but not with respect to another. Making a distinction between Θ and α allows us to highlight the specific explanation (which we did not have before), and control different forms of explanation (facts, rules, or even new theories). But certainly, our accounts would become yet more sensitive if we worked with some of the above richer formats.

Inferential Strength: A Parameter

At first glance, once we have Tarski's notion of truth, logical consequence seems an obvious defined notion. A conclusion follows if it is true in all models where the premisses are true. But the contemporary philosophical and computational traditions have shown that natural notions of inference may need more than truth in the above sense, or may even hinge on different properties altogether. For example, among the candidates that revolve around truth, statistical inference requires not total inclusion of premisse models in conclusion models, but only a significant overlap, resulting in a high degree of certainty. Other approaches introduce new semantic primitives. Notably, Shoham's notion of causal and default reasoning ([Sho88]) introduces a preference order on models, requiring only that the *most preferred models* of Σ be included in the models of φ.

More radically, dynamic semantics replaces the notion of truth by that of *information change*, aiming to model the flow of information. This move leads to a redesign for Tarski semantics, with e.g. quantifiers becoming actions on assignments ([vBC94]). This logical paradigm has room for many different inferential notions ([Gro95, vBe96a]). An example is the earlier mentioned update-to-test-consequence:

"process the successive premisses in Σ, thereby absorbing their informational content into the initial information state. At the end, check if the resulting state is rich enough to satisfy the conclusion φ".

Informational content rather than truth is also the key semantic property in situation theory ([PB83]). In addition to truth-based and information-based approaches, there are, of course, also various proof-theoretic variations on standard consequence. Examples are default reasoning: "φ is provable unless and until φ is disproved" ([Rei80]), and indeed Hempel's hypothetico-deductive model of scientific inference itself.

All these alternatives agree with our analysis of abductive explanatory inference. On our view, abduction is not a new notion of inference. It is rather a topic-dependent practice of scientific reasoning, which can be supported by various members of the above family. In fact, it is appealing to think of abductive inference in several respects, as inference involving preservation of both truth and *explanatory power*. In fact, appropriately defined, both might turn out equivalent. It has also been argued that since abduction is a form of reversed deduction, just as deduction is truth-preserving, abduction must be falsity-preserving ([Mic94]). However, [Fla95] gives convincing arguments against this particular move. Moreover, as we have already discussed intuitively, abduction is not just deduction in reverse.

Our choice here is to study abductive inference in more depth as a strengthened form of classical inference. This is relevant, it offers nice connections with artificial intelligence and the philosophy of science, and it gives a useful simple start for a broader systematic study of abductive inference. One can place this choice in a historical context, namely the work of Bernard Bolzano, a nineteenth century philosopher and mathematician (and theologian) engaged in the study of different varieties of inference. We provide a brief excursion, providing some perspective for our later technical considerations.

Bolzanos's Program

Bolzano's notion of deducibility (*Ableitbarkeit*) has long been recognized as a predecessor of Tarski's notion of logical consequence ([Cor75]). However, the two differ in several respects, and in our broader view of logic, they even appear radically different. These differences have been studied both from a philosophical ([Tho81]) and from a logical point of view ([vBe84a]).

One of Bolzano's goals in his theory of science ([Bol73]), was to show why the claims of science form a theory as opposed to an arbitrary set of propositions. For this purpose, he defines his notion of deducibility as a logical relationship extracting conclusions from premisses forming *compatible propositions*, those for which some set of ideas make all propositions true when uniformly substituted throughout. In addition, compatible propositions must share *common ideas*. Bolzano's use of 'substitutions' is of interest by itself, but for our purposes here, we will identify these (somewhat roughly) with the standard use of 'models'. Thompson attributes the difference between Bolzano's consequence and Tarski's to the fact that the former notion is epistemic while the latter is ontological. These differences have strong technical effects. With Bolzano, the premisses must be consistent (sharing at least one model), with Tarski, they need not. Therefore, from a contradiction, everything follows for Tarski, and nothing for Bolzano.

Restated for our ternary format, then, Bolzano's notion of deducibility reads as follows (cf. [vBe84a]):

$\Theta \mid \alpha \Rightarrow \varphi$ if
(1) The conjunction of Θ and α is consistent.
(2) Every model for Θ plus α verifies φ.

Therefore, Bolzano's notion may be seen (anachronistically) as Tarski's consequence plus the additional condition of consistency. Bolzano does not stop here. A finer grain to deducibility occurs in his notion of *exact deducibility*, which imposes greater requirements of 'relevance'. A modern version, involving inclusion-minimality for sets of abducibles, may be transcribed (again, with some historical injustice) as:

$\Theta \mid \alpha \Rightarrow^{+} \varphi$ if
(1) $\Theta \mid \alpha \Rightarrow \varphi$
(2) There is no proper subset of α, α', such that $\Theta \mid \alpha' \Rightarrow \varphi$.

That is, in addition to consistency with the background theory, the premisse set α must be 'fully explanatory' in that no subpart of it would do the derivation. Notice that this leads to non-monotonicity. Here is an example:

$\Theta \mid a \rightarrow b, a \Rightarrow^{+} b$
$\Theta \mid a \rightarrow b, a, b \rightarrow c \not\Rightarrow^{+} b$

Bolzano's agenda for logic is relevant to our study of abductive reasoning (and the study of general non-monotonic consequence relations) for several reasons. It suggests the methodological point that what we need is not so much proliferation of different logics as a better grasp of different styles of consequence. Moreover, his work reinforces an earlier claim, that truth is not all there is to understanding explanatory reasoning. More specifically, his notions still have interest. For example, exact deducibility has striking similarities to explanation in philosophy of science (cf. chapter 5).

Abductive Explanatory Inference as Deduction in Reverse

In this section we define abductive explanatory inference as a strengthened form of classical inference. Our proposal will be in line with abduction in artificial intelligence, as well as with the Hempelian account of explanation. We will motivate our requirements with our very simple rain example, presented here in classical propositional logic:

$\Theta : r \rightarrow w, s \rightarrow w$
$\varphi : w$

The first condition for a formula α to count as an explanation for φ with respect to Θ is the inference requirement. Many formulas would satisfy this condition. In addition to earlier-mentioned obvious explanations (r: rain, s: sprinklers-on), one might take their conjunction with any other formula, even if the latter is inconsistent with Θ (e.g. $r \wedge \neg w$). One can take the fact itself (w), or, one can introduce entirely new facts and rules (say, there are children playing with water, and this causes the lawn to get wet).

Inference: $\Theta, \alpha \models \varphi$
α's: $r, s, r \wedge s, r \wedge z, r \wedge \neg w, s \wedge \neg w, w, [c, c \rightarrow w], \Theta \rightarrow w$.

Some of these 'abductive explanations' must be ruled out from the start. We therefore impose a consistency requirement on the left hand side, leaving only the following as possible abductive explanations:

Consistency: Θ, α is consistent.
α's: $r, s, r \wedge s, r \wedge z, w, [c, c \rightarrow w], \Theta \rightarrow w$.

An abductive explanation α is only *necessary*, if φ is not already entailed by Θ. Otherwise, any consistent formula will count as an abductive explanation. Thus we repeat an earlier trigger for abduction: $\Theta \not\models \varphi$. By itself, this does not rule out any potential abducibles on the above list (as it does not involve the argument α.) But also, in order to avoid what we may call *external explanations* –those that do not use the background theory at all (like the explanation involving children in our example) –, it must be required that α be insufficient for explaining φ by itself ($\alpha \not\models \varphi$). In particular this condition avoids the trivial reflexive explanation $\varphi \not\Rightarrow \varphi$. Then only the following explanations are left in our list of examples:

Explanation $\Theta \not\models \varphi, \alpha \not\models \varphi$
α's: $r, s, r \wedge s, r \wedge z, \Theta \rightarrow w$.

Now both Θ and α contribute to explaining φ. However, we are still left with some formulas that do not seem to be genuine explanations ($r \wedge z, \Theta \rightarrow w$). Therefore, we explore a more sensitive criterion, admitting only 'the best explanation'.

Selecting the Best Explanation

Intuitively, a reasonable ground for choosing a statement as the best explanation, is its simplicity. It should be minimal, i.e. as weak as possible in performing its job. This would lead us to prefer r over $r \wedge z$ in the preceding example. As Peirce puts it, we want the explanation that *"adds least to what has been observed"* (cf. [CP, 6.479]). The criterion of simplicity has been extensively considered both in the philosophy of science and in artificial intelligence. But its precise formulation remains controversial, as measuring simplicity can be a tricky matter. One attempt to capture simplicity in a logical way is as follows:

Weakest Abductive Explanation:
α is the weakest abductive explanation for φ with respect to Θ iff

(i) $\Theta, \alpha \models \varphi$

(ii) For all other formulas β such that $\Theta, \beta \models \varphi$, $\models \beta \rightarrow \alpha$.

This definition makes the explanations r and s almost the weakest in the above example, just as we want. Almost, but not quite. For, the explanation $\Theta \rightarrow w$, a trivial solution, turns out to be the minimal one. The following is a folklore observation to this effect:

FACT 1 *Given any theory Θ and observation φ to be explained from it, $\alpha = \Theta \rightarrow \varphi$ is the weakest abductive explanation.*

Proof. Obviously, we have (i) $\Theta, \Theta \rightarrow \varphi \models \varphi$. Moreover, let α' be any other explanation. This says that $\Theta, \alpha' \models \varphi$. But then we also have (by conditionalizing) that $\alpha' \models \Theta \rightarrow \varphi$, and hence $\models \alpha' \rightarrow (\Theta \rightarrow \varphi)$ $\dashv$

That $\Theta \rightarrow \varphi$ is a solution that will always count as an explanation in a deductive format was noticed by several philosophers of science ([Car55]). It has been used as an argument to show how the issue would impose restrictions on the syntactic form of abducibles. Surely, in this case, the explanation seems too complex to count. We will therefore reject this proposal, noting also that it fails to recognize (let alone compare) intuitively 'minimal' explanations like r and s in our running example.

Other criteria of minimality exist in the literature. One of them is based on preference orderings. The best explanation is the most preferred one, given an explicit ordering of available assertions. In our example, we could define an order in which inconsistent explanations are the least preferred, and the simplest the most. These preference approaches are quite flexible, and can accommodate various working intuitions. However, they may still depend on many factors, including the background theory. This seems to fall outside a logical framework, referring rather to further 'economic' decision criteria like utilities. A case in point is Peirce's 'economy of research' in selecting a most promising hypothesis. What makes a hypothesis good or best has no easy answer. One may appeal to criteria of simplicity, likelihood, or predictive power. To complicate matters even further, we often do not compare (locally) quality of explanations given a fixed theory, but rather (globally) whole packages of 'theory + explanation'. This perspective gives a much greater space of options. As we have not been able to shed a new light from logic upon these matters, we will ignore these dimensions here.

Further study would require more refined views of theory structure and reasoning practice, in line with some of the earlier references[7], or even more ambitiously, following current approaches to 'verisimilitude' in the philosophy of science (cf. [Kui87]).

[7] Preferences over models (though not over statements) will be mentioned briefly as providing one possible inference mechanism for abduction.

We conclude with one final observation. perhaps one reason why the notion of 'minimality' has proved so elusive is again our earlier product-process distinction. Philosophers have tried to define minimality in terms of intrinsic properties of statements and inferences as products. But it may rather be a process-feature, having to do with computational effort in some particular procedure performing abduction. Thus, one and the same statement might be minimal in one abduction, and non-minimal in another.

Abductive Explanatory Characterization Styles

Following our presentation of various requirements for an abductive explanation, we make things more concrete for further reference. We consider five versions of abductive explanations making up the following styles: plain, consistent, explanatory, minimal and preferential, defined as follows:

Abductive Explanatory Styles

Given Θ (a set of formulae) and φ (a sentence), α is an *abductive explanation* if:

Plain :

(i) $\Theta, \alpha \models \varphi$.

Consistent :

(i) $\Theta, \alpha \models \varphi$,

(ii) Θ, α consistent.

Explanatory :

(i) $\Theta, \alpha \models \varphi$,

(ii) $\Theta \not\models \varphi$,

(iii) $\alpha \not\models \varphi$.

Minimal :

(i) $\Theta, \alpha \models \varphi$,

(ii) α is the weakest such abductive explanation (not equal to $\Theta \rightarrow \varphi$).

Preferential :

(i) $\Theta, \alpha \models \varphi$,

(ii) α is the best abductive explanation according to some given preferential ordering.

We can form other combinations, of course, but these will already exhibit many characteristic phenomena. Note that these requirements do not depend on classical consequence. For instance, in Chapter 5, the consistency and the explanatory requirements work just as well for statistical inference. The former then also concerns the explanandum φ. (For, in probabilistic reasoning it is possible to infer two contradictory conclusions even when the premisses are consistent.) The latter helps capture when an explanation helps raise the probability of the explanandum.

A full version of abduction would make the formula to be abduced part of the derivation, consistent, explanatory, and the best possible one. However, instead of incorporating all these conditions at once, we shall consider them one by one. Doing so clarifies the kind of restriction each requirement adds to the notion of plain abduction. Our standard versions will base these requirements on classical consequence underneath. But we also look briefly toward the end at versions involving other notions of consequence. We will find that our various notions of abduction have advantages, but also drawbacks, such as an increase of complexity for explanatory reasoning as compared with classical inference.

Up to now, we can say that abductive inference may be characterized by reversed deduction plus additional conditions. However, is this all we can say about the logic of abduction? This definition does not really capture the rationality principles behind this type of reasoning, like its non-monotonic feature we have talked about.

In what follows, our aim is to present the characterization of abductive inference from the structural perspective we introduced earlier in this chapter. This approach has become popular across a range of non-standard logics. Our systematic analysis will explore different abductive styles from this perspective.

4. Abductive Explanatory Inference: Structural Characterization

Consistent Abductive Explanatory Inference

We recall the definition:

$\Theta \mid \alpha \Rightarrow \varphi$ iff
(i) $\Theta, \alpha \models \varphi$
(ii) Θ, α are consistent

The first thing to notice is that the two items to the left behave *symmetrically*:

$$\Theta \mid \alpha \Rightarrow \varphi \qquad \text{iff} \qquad \alpha \mid \Theta \Rightarrow \varphi$$

Indeed, in this case, we may technically simplify matters to a binary format after all: $X \Rightarrow C$, in which X stands for the conjunction of Θ and α, and C for φ. To bring these in line with the earlier-mentioned structural analysis of nonclassical logics, we view X as a finite sequence $X_1 \ldots, X_k$ of formulas and C as a single conclusion.

Classical Structural Rules

Of the structural rules for classical consequence, contraction and permutation hold for consistent abduction. But reflexivity, monotonicity and cut fail, witness by the following counterexamples:

- Reflexivity: $p \wedge \neg p \not\Rightarrow p \wedge \neg p$
- Monotonicity: $p \Rightarrow p$, but $p, \neg p \not\Rightarrow p$
- Cut: $p, \neg q \Rightarrow p$, and $p, q \Rightarrow q$, but $p, \neg q, q \not\Rightarrow q$

New Structural Rules

Here are some restricted versions of the above failed rules, and some others that are valid for consistent abduction:

1 Conditional Reflexivity (**CR**)

$$\frac{X \Rightarrow B}{X \Rightarrow X_i} \qquad 1 \leq i \leq k$$

2 Simultaneous Cut (**SC**)

$$\frac{U \Rightarrow A_1 \ldots U \Rightarrow A_k \qquad A_1, \ldots, A_k \Rightarrow B}{U \Rightarrow B}$$

3 Conclusion Consistency (**CC**)

$$\frac{U \Rightarrow A_1 \ldots U \Rightarrow A_k}{A_1, \ldots, A_k \Rightarrow A_i} \qquad 1 \leq i \leq k$$

These rules state the following. Conditional Reflexivity requires that the sequence X derive something else ($X \Rightarrow B$), as this ensures consistency. Simultaneous Cut is a combination of Cut and Contraction in which the sequent $A_1, \ldots, A_k$ may be omitted in the conclusion when each of its elements A_i is consistently derived by U and this one in its turn consistently derives B. Conclusion Consistency says that a sequent $A_1, \ldots, A_k$ implies its elements if each of these are implied consistently by something (U arbitrary), which is another form of reflexivity.

PROPOSITION 2 *These rules are sound for consistent abduction.*

Proof. In each of these three cases, it is easy to check by simple set-theoretic reasoning that the corresponding classical consequence holds. Therefore, the only thing to be checked is that the premisses mentioned in the conclusions of these rules must be consistent. For Conditional Reflexivity, this is because X already consistently implied something. For Simultaneous Cut, this is because U already consistently implied something. Finally, for Conclusion Consistency, the reason is that U must be consistent, and it is contained in the intersection of all the A_i, which is therefore consistent, too. ⊣

A Representation Theorem

The given structural rules in fact characterize consistent abduction:

PROPOSITION 3 *A consequence relation satisfies structural rules 1 (CR), 2 (SC), 3 (CC) iff it is representable in the form of consistent abduction.*

Proof. Soundness of the rules was proved above. Now consider the completeness direction. Let $\Rightarrow$ be any abstract relation satisfying 1, 2, 3. Define for any proposition A,

$$A^* = \{X \mid X \Rightarrow A\}$$

We now show the following statement of adequacy for this representation:

Claim. $A_1, \ldots, A_k \Rightarrow B$ iff $\emptyset \subset A_1^* \cap \ldots \cap A_k^* \subseteq B^*$.

Proof. 'Only if'. Since $A_1, \ldots, A_k \Rightarrow B$, by Rule 1 (CR) we have $A_1, \ldots, A_k \Rightarrow A_i$ $(1 \leq i \leq k)$. Therefore, $A_1, \ldots, A_k \in A_i^*$, for each i with $1 \leq i \leq k$, which gives the proper inclusion. Next, let U be any sequence in the intersection of all A_i^*, for $1, \ldots, k$. That is, $U \Rightarrow A_1, \ldots, U \Rightarrow A_k$. By Rule 2 (SC), $U \Rightarrow B$, i.e. $U \in B^*$, and we have shown the second inclusion.

'If'. Using the assumption of non-emptiness, let, say, $U \in \bigcap A_i^*$, for $1, \ldots, k$. i.e. $U \Rightarrow A_1, \ldots, U \Rightarrow A_k$. By Rule 3 (CC), $A_1, \ldots, A_k \Rightarrow A_i$ $(1 \leq i \leq k)$. By the second inclusion then, $A_1, \ldots, A_k \in B^*$. By the definition of the function *, this means that $A_1, \ldots, A_k \Rightarrow B$. ⊣

More Familiar Structural Rules

The above principles characterize consistent abduction. Even so, there are more familiar structural rules that are valid as well, including modified forms of Monotonicity and Cut. For instance, it is easy to see that $\Rightarrow$ satisfies a form of modified monotonicity: B may be added as a premisse if this addition does not endanger consistency. And the latter may be shown by their 'implying' any conclusion:

- Modified Monotonicity:

$$\frac{X \Rightarrow A \qquad X, B \Rightarrow C}{X, B \Rightarrow A}$$

As this was not part of the above list, we expect some derivation from the above principles. And indeed there exists one:

- Modified Monotonicity Derivation:

$$\cfrac{\cfrac{X, B \Rightarrow C}{X, B \Rightarrow X_i^{\prime}s}\,1 \qquad X \Rightarrow A}{X, B \Rightarrow A}\,2$$

These derivations also help in seeing how one can reason perfectly well with non-classical structural rules. Another example is the following valid form of Modified Cut:

- Modified Cut

$$\frac{X \Rightarrow A \qquad U, A, V \Rightarrow B \qquad U, X, V \Rightarrow C}{U, X, V \Rightarrow B}$$

This may be derived as follows:

- Modified Cut Derivation

$$\dfrac{\dfrac{U,X,V \Rightarrow C}{U,X,V \Rightarrow U's,V's}\,{}^{1} \qquad \dfrac{\dfrac{U,X,V \Rightarrow C}{U,X,V \Rightarrow X_i's}\,{}^{1} \qquad X \Rightarrow A}{U,X,V \Rightarrow A}\,{}^{2} \qquad U,A,V \Rightarrow B}{U,X,V \Rightarrow B}\,{}^{2}$$

Finally, we check some classically structural rules that do remain valid as they stand, showing the power of Rule (3):

- Permutation

$$\dfrac{\dfrac{\dfrac{X,A,B,Y \Rightarrow C}{X,A,B,Y \Rightarrow X \quad X,A,B,Y \Rightarrow A \quad X,A,B,Y \Rightarrow B \quad X,A,B,Y \Rightarrow Y}\,{}^{1}}{X,B,A,Y \Rightarrow X \quad X,B,A,Y \Rightarrow B \quad X,B,A,Y \Rightarrow A \quad X,B,A,Y \Rightarrow Y}\,{}^{3} \qquad X,A,B,Y \Rightarrow C}{X,B,A,Y \Rightarrow C}\,{}^{2}$$

- Contraction (one sample case)

$$\dfrac{\dfrac{\dfrac{X,A,A,Y \Rightarrow B}{X,A,A,Y \Rightarrow X_i's, A, Y_i's}\,{}^{1}}{X,A,Y \Rightarrow X_i's, A, Y_i's}\,{}^{3} \qquad X,A,A,Y \Rightarrow B}{X,A,Y \Rightarrow B}\,{}^{2}$$

Thus, consistent abductive inference defined as classical consequence plus the consistency requirement has appropriate forms of reflexivity, monotonicity, and cut for which it is assured that the premisses remain consistent. Permutation and contraction are not affected by the consistency requirement, therefore the classical forms remain valid. More generally, the preceding examples show simple ways of modifying all classical structural principles by putting in one extra premisse ensuring consistency.

Simple as it is, our characterization of this notion of inference does provide a complete structural description of Bolzano's notion of deducibility introduced earlier in this chapter (section 3.3).

Explanatory Abductive (Explanatory) Inference

Explanatory abductive explanatory inference (explanatory abduction, for short)was defined as plain abduction ($\Theta, \alpha \models \varphi$) plus two conditions of necessity ($\Theta \not\models \varphi$) and insufficiency ($\alpha \not\models \varphi$). However, we will consider a weaker version (which only considers the former condition) and analyze its structural rules. This is actually somewhat easier from a technical viewpoint. The full version remains of general interest though, as it describes the 'necessary collaboration' of two premisses set to achieve a conclusion. It will be analyzed further in chapter 5 in connection with philosophical models of scientific explanation. We rephrase our notion as:

Weak Explanatory Abduction:
$\Theta \mid \alpha \Rightarrow \varphi$ iff
(i) $\Theta, \alpha \models \varphi$
(ii) $\Theta \not\models \varphi$

The first thing to notice is that we must leave the binary format of premisses and conclusion. This notion is non-symmetric, as Θ and α have different roles. Given such a ternary format, we need a more finely grained view of structural rules. For instance, there are now two kinds of monotonicity, one when a formula is added to the explanations and the other one when it is added to the theory:

- Monotonicity for Abductive Explanations:

$$\frac{\Theta \mid \alpha \Rightarrow \varphi}{\Theta \mid \alpha, A \Rightarrow \varphi}$$

- Monotonicity for Theories:

$$\frac{\Theta \mid \alpha \Rightarrow \varphi}{\Theta, A \mid \alpha \Rightarrow \varphi}$$

The former is valid, but the latter is not. (A counterexample is: $p \mid q, r \Rightarrow q$ but $p, q \mid q, r \not\Rightarrow q$). Monotonicity for explanations states that an explanation for a fact does not get invalidated when we strengthen it, as long as the theory is not modified.

Here are some valid principles for weak explanatory abduction.

- Weak Explanatory Reflexivity

$$\frac{\Theta \mid \alpha \Rightarrow \varphi}{\Theta \mid \varphi \Rightarrow \varphi}$$

- Weak Explanatory Cut

$$\frac{\Theta \mid \alpha, \beta \Rightarrow \varphi \qquad \Theta \mid \alpha \Rightarrow \beta}{\Theta \mid \alpha \Rightarrow \varphi}$$

In addition, the classical forms of contraction and permutation are valid on each side of the bar. Of course, one should not permute elements of the theory with those in the explanation slot, or vice versa. We conjecture that the given principles completely characterize the weak explanatory abduction notion, when used together with the above valid form of monotonicity.

Structural Rules with Connectives

Pure structural rules involve no logical connectives. Nevertheless, there are natural connectives that may be used in the setting of abductive consequence. For instance, all Boolean operations can be used in their standard meaning. These, too, will give rise to valid principles of inference. In particular, the following well-known classical laws hold for all notions of abductive inference studied so far:

- Disjunction of Θ-antecedents:

$$\frac{\Theta_1 \mid A \Rightarrow \varphi \qquad \Theta_2 \mid A \Rightarrow \varphi}{\Theta_1 \vee \Theta_2 \mid A \Rightarrow \varphi}$$

- Conjunction of Consequents

$$\frac{\Theta \mid A \Rightarrow \varphi_1 \qquad \Theta \mid A \Rightarrow \varphi_2}{\Theta \mid A \Rightarrow \varphi_1 \wedge \varphi_2}$$

These rules will play a role in our proposed calculus for abduction, as we will show later on.

Another way of expressing monotonicity with the aid of negation and classical derivability is as follows:

- Monotonicity:

$$\frac{\Theta \mid \alpha \Rightarrow \varphi \qquad \Theta \mid \alpha \not\vdash \neg\beta}{\Theta \mid \alpha, \beta \Rightarrow \varphi}$$

We conclude a few brief points on the other versions of abduction on our list. We have not undertaken to characterize these in any technical sense.

Minimal and Preferential Abductive Explanatory Inference

Consider our versions of 'minimal' abduction. One said that $\Theta, \alpha \models \varphi$ and α is the weakest such explanation. By contrast, preferential abduction said that $\Theta, \alpha \models \varphi$ and α is the best explanation according to some given preferential ordering. For the former, with the exception of the above disjunction rule for antecedents, no other rule that we have seen is valid. But it does satisfy the following form of transitivity:

- Transitivity for Minimal Abduction:

$$\frac{\Theta \mid \alpha \Rightarrow \varphi \qquad \Theta \mid \beta \Rightarrow \alpha}{\Theta \mid \beta \Rightarrow \varphi}$$

For preferential abduction, on the other hand, no structural rule formulated so far is valid. The reason is that the relevant preference order amongst formulas, in itself needs to be captured in the formulation of our inference rules. A valid formulation of monotonicity would then be something along the following lines:

- Monotonicity for Preferential Abduction:

$$\frac{\Theta \mid \alpha \Rightarrow \varphi \qquad \alpha, \beta < \alpha}{\Theta \mid \alpha, \beta \Rightarrow \varphi}$$

In our opinion, this is no longer a structural rule, since it adds a mathematical relation ($<$ for a preferential order) that cannot in general be expressed in terms of the consequence itself. This is a point of debate, however, and its solution depends on what each logic artisan is willing to represent in a logic. In any case, this format is beyond what we will study in this book.

Structural Rules for Nonstandard Inference

All abductive versions so far had classical consequence underneath. In this section, we briefly explore structural behaviour when the underlying notion of inference is non standard, as in preferential entailment. Moreover, we throw in some words about structural rules for abduction in logic programming, and for induction.

Preferential Reasoning

Interpreting the inferential parameter as preferential entailment means that $\Theta, \alpha \Rightarrow \varphi$ if (only) the most preferred models of $\Theta \cup \alpha$ are included in the models of φ. This leads to a completely different set of structural rules. Here are some valid examples, transcribed into our ternary format from [KLM90]:

- Reflexivity: $\Theta, \alpha \Rightarrow \alpha$
- Cautious Monotonicity:

$$\frac{\Theta \mid \alpha \Rightarrow \beta \qquad \Theta \mid \alpha \Rightarrow \gamma}{\Theta \mid \alpha, \beta \Rightarrow \gamma}$$

- Cut:

$$\frac{\Theta \mid \alpha, \beta \Rightarrow \gamma \qquad \alpha \Rightarrow \beta}{\Theta \mid \alpha \Rightarrow \gamma}$$

- Disjunction:

$$\frac{\Theta \mid \alpha \Rightarrow \varphi \qquad \Theta \mid \beta \Rightarrow \varphi}{\Theta \mid \alpha \vee \beta \Rightarrow \varphi}$$

It is interesting to see in greater detail what happens to these rules when we add our further conditions of 'consistency' and 'explanation'. In all, what happens is merely that we get structural modifications similar to those found earlier on for classical consequence. Thus, a choice for a preferential proof engine, rather than classical consequence, seems orthogonal to the behavior of abduction.

Structural rules for Prolog Computation

An analysis via structural rules may be also performed for notions of $\Rightarrow$ with a more procedural flavor. In particular, the earlier-mentioned case of Prolog computation obeys clear structural rules (cf. [vBe92, Kal95, Min90]). Their format is somewhat different from classical ones, as one needs to represent more of the Prolog program structure for premisses, including information on rule heads. (Also, Kalsbeek [Kal95] gives a complete calculus of structural rules for logic programming including such control devices as the cut operator !). The characteristic expressions of a Gentzen style sequent calculus for these systems (in the reference above) are sequents of the form $[P] \Rightarrow \varphi$, where P is a (propositional, Horn clause) program and φ is an atom. A failure of a goal is expressed as $[P] \Rightarrow \neg\varphi$ (meaning that φ finitely fails). In this case, valid monotonicity rules must take account of the place in which premisses are added, as Prolog is sensitive to the order of its program clauses. Thus, of the following rules, the first one is valid, but the second one is not:

- Right Monotonicity

$$\frac{[P] \Rightarrow \varphi}{[P; \beta] \Rightarrow \varphi}$$

- Left Monotonicity

$$\frac{[P] \Rightarrow \varphi}{[\beta; P] \Rightarrow \varphi}$$

Counterexample: $\beta = \varphi \leftarrow \varphi$

The question of complete structural calculi for abductive logic programming will not be addressed in this book, we will just mention that a natural rule for an 'abductive update' is as follows:

- Atomic Abductive Update

$$\frac{[P] \Rightarrow \neg\varphi}{[P; \varphi] \Rightarrow \varphi}$$

We will briefly return to structural rules for abduction as a process in the next chapter.

Structural Rules For Induction

Unlike abduction, enumerative induction is a type of inference that explains a set of observations, and makes a prediction for further ones (cf. our discussion in chapter 2). Our previous rule for conjunction of consequents already suggests how to give an account for further observations, provided that we interpret the commas below as conjunction amongst formulae (in the usual Gentzen calculus, commas to the right are interpreted rather as disjunctions):

$$\frac{\alpha \Rightarrow \varphi_1 \qquad \alpha \Rightarrow \varphi_2}{\alpha \Rightarrow \varphi_1, \varphi_2}$$

That is, an inductive explanation α for φ_1 remains an explanation when a formula φ_2 is added, provided that α also accounts for it separately. Note that this rule is a kind of monotonicity, but this time the increase is on the conclusion set rather than on the premisse set. More generally, an inductive explanation α for a set of formulae remains valid for more input data ψ when it explains it:

- (Inductive) Monotonicity on Observations

$$\frac{\Theta \mid \alpha \Rightarrow \varphi_1, \ldots, \varphi_n \qquad \Theta \mid \alpha \Rightarrow \psi}{\Theta \mid \alpha \Rightarrow \varphi_1, \ldots, \varphi_n, \psi}$$

In order to put forward a set of rules characterizing inductive explanation, a further analysis of its properties should be made, and this falls beyond the scope of this thesis. What we anticipate however, is that a study of enumerative induction from a structural point of view will bring yet another twist to the standard structural analysis, that of giving an account of changes in conclusions.

Further Logical Issues

Our analysis so far has only scratched the surface of a broader field. In this section we discuss a number of more technical logical aspects of abductive styles of inference. This identifies further issues that seem relevant to understanding the logical properties of abduction.

Completeness

The usual completeness theorems have the following form:

$\Theta \models \varphi \qquad \text{iff} \qquad \Theta \vdash \varphi$

With our ternary format, we would expect some similar equivalence, with a possibly different treatment of premisses on different sides of the comma:

$\Theta, \alpha \models \varphi \qquad \text{iff} \qquad \Theta, \alpha \vdash \varphi$

Can we get such completeness results for any of the abductive versions we have described so far? Here are two extremes.

The representation arguments for the above characterizations of abduction may be reworked into completeness theorems of a very simple kind. (This works just as in [vBe96a], chapter 7). In particular, for consistent abduction, our earlier argument essentially shows that $\Theta, \alpha \Rightarrow \varphi$ follows from a set of ternary sequents Φ iff it can be derived from Φ using only the derivation rules (CR), (SC), (CC) above.

These representation arguments may be viewed as 'poor man's completeness proofs', for a language without logical operators. Richer languages arise by adding operators, and completeness arguments need corresponding 'upgrading' of the representations used. (Cf. [Kur95] for an elaborate analysis of this upward route for the case of categorial and relevance logics. [Gro95] considers the same issue in detail for dynamic styles of inference.) At some level, no more completeness theorems are to be expected. The complexity of the desired proof theoretical notion $\vdash$ will usually be recursively enumerable (Σ^0_1). But, our later analysis will show that, with a predicate-logical language, the complexity of semantic abduction $\models$ will become higher than that. The reason is that it mixes derivability with non-derivability (because of the consistency condition).

So, our best chance for achieving significant completeness is with an intermediate language, like that of propositional logic. In that case, abduction is still decidable, and we may hope to find simple proof rules for it as well. (Cf. [Tam94] for the technically similar enterprise of completely axiomatizing simultaneous 'proofs' and 'fallacies' in propositional logic.) Can we convert our representation arguments into full-fledged completeness proofs when we add propositional operators $\neg, \wedge, \vee$? We have already seen that we do get natural valid principles like disjunction of antecedents and conjunction of consequents. However, there is no general method that connects a representational result into more familiar propositional completeness arguments. A case of successful (though non-trivial) transfer is in [Kan93], but essential difficulties are identified in [Gro95].

Instead of solving the issue of completeness here, we merely propose the following axioms and rules for a sequent calculus for consistent abduction (which we label as $\models_c$) in what follows:

- Axiom: $p \models_c p$
- Rules for Conjunction:

$$\wedge_1 \frac{\Theta \models_c \varphi_1, \qquad \Theta \models_c \varphi_2}{\Theta \models_c \varphi_1 \wedge \varphi_2}$$

The following are valid provided that α, ψ are formulas with only positive propositional letters:

$$\wedge_2 \frac{\alpha \models_c \alpha \qquad \psi \models_c \psi}{\alpha, \psi \models_c \alpha}$$

$$\wedge_3 \frac{\alpha, \psi \models_c \varphi}{\alpha \wedge \psi \models_c \varphi}$$

- Rules For Disjunction:

$$\vee_1 \frac{\Theta_1 \models_c \varphi \qquad \Theta_2 \models_c \varphi}{\Theta_1 \vee \Theta_2 \models_c \varphi}$$

$$\vee_2 \frac{\Theta \models_c \varphi}{\Theta \models_c \varphi \vee \psi}$$

$$\vee_3 \frac{\Theta \models_c \varphi}{\Theta \models_c \psi \vee \varphi}$$

- Rules for Negation:

$$\neg_1 \frac{\Theta, A \models_c \varphi}{\Theta \models_c \varphi \vee \neg A}$$

$$\neg_2 \frac{\Theta \models_c \varphi \vee A \qquad \Theta \wedge \neg A \models_c \psi}{\Theta \wedge \neg A \models_c \varphi}$$

It is easy to see that these rules are sound on the interpretation of $\models$ as consistent abduction. This calculus is already unlike most usual logical systems, though. First of all there is no substitution rule, as $p \models p$ is an axiom, whereas in general $\psi \not\models \psi$ unless ψ has only positive propositional letters, in which case it is proved to be consistent. By itself, this is not dramatic (for instance, several modal logics exist without a valid substitution rule), but it is certainly uncommon. Moreover, note that the rules which "move things to the left" ($\neg_2$) are different from their classical counterparts, and others ($\wedge_3$) are familiar but here a condition to ensure consistency is added. Even so, one can certainly do practical work with a calculus like this.

For instance, all valid principles of classical propositional logic that do not involve negations are derivable here. Semantically, this makes sense, as positive formulas are always consistent without special precautions. On the other hand, it is easy to check that the calculus provides no proof for a typically invalid sequent like $p \wedge \neg p \models p \wedge \neg p$[8].

[8] The reason is that their cut-free classical proofs (satisfying the subformula property) involve only conjunction and disjunction - for which we have the standard rules.

Digression:
A general semantic view of abductive explanatory consequence

Speaking generally, we can view a ternary inference relation $\Theta \mid \alpha \Rightarrow \varphi$ as a ternary relation C (T, A, F) between sets of models for, respectively, Θ, α, and φ. What structural rules do is constrain these relations to just a subclass of all possibilities. (This type of analysis has analogies with the theory of generalized quantifiers in natural language semantics. It may be found in [vBe84a] on the model theory of verisimilitude, or in [vBe96b] on general consequence relations in the philosophy of science.) When enough rules are imposed we may represent a consequence relation by means of simpler notions, involving only part of the a priori relevant $2^3 = 8$ "regions" of models induced by our three argument sets.

In this light, the earlier representation arguments might even be enhanced by including logical operators. We merely provide an indication. It can be seen easily that, in the presence of disjunction, our explanatory abduction satisfies full Boolean 'Distributivity' for its abducible argument α_i:

$$\Theta \mid \bigvee_i \alpha_i \Rightarrow \varphi \qquad \text{iff} \qquad \text{for some i, } \Theta \mid \alpha_i \Rightarrow \varphi.$$

Principles like this can be used to reduce the complexity of a consequence relation. For instance, the predicate argument A may now be reduced to a point wise one, as any set A is the union of all singletons $\{a\}$ with $a \in$ A.

Complexity

Our next question addresses the complexity of different versions of abduction. Non-monotonic logics may be better than classical ones for modelling common sense reasoning and scientific inquiry. But their gain in expressive power usually comes at the price of higher complexity, and abduction is no exception. Our interest is then to briefly compare the complexity of abduction to that of classical logic. We have no definite results here, but we do have some conjectures. In particular, we look at consistent abduction, beginning with predicate logic.

Predicate-logical validity is undecidable by Church's Theorem. Its exact complexity is Σ^0_1 (the validities are recursively enumerable, but not recursive). (To understand this outcome, think of the equivalent assertion of derivability: "there exists a P: P is a proof for φ".) More generally, Σ (or Π) notation refers to the usual prenex forms for definability of notions in the Arithmetical Hierarchy. Complexity is measured here by looking at the quantifier prenex, followed by a decidable matrix predicate. A subscript n indicates n quantifier changes in the prenex. (If a notion is both Σ_n and Π_n, it is called Δ_n.) The complementary notion of satisfiability is also undecidable, being definable in the form Π^0_1. Now, abductive consequence moves further up in this hierarchy.

In order to show that consistent abduction is not Δ^0_2-complete we have the following. The statement that "Θ, α is consistent" is Π^0_1, while the statement that

"$\Theta, \alpha \models \varphi$" is Σ_1^0 (cf. the above observations). Therefore, their conjunction may be written, using well-known prenex operations, in either of the following forms:

$$\exists\forall DEC \qquad \text{or} \qquad \forall\exists DEC.$$

Hence consistent abduction is in Δ_2^0. This analysis gives an upper bound only. But we cannot do better than this. So it is also a lower bound. For the sake of reductio, suppose that consistent abduction were Σ_1^0. Then we could reduce satisfiability of any formula B effectively to the abductive consequence $B, B \Rightarrow B$, and hence we would have that satisfiability is also Σ_1^0. But then, Post's Theorem says that a notion which is both Σ_1^0 and Π_1^0 must be decidable. This is a contradiction, and hence $\Theta, \alpha \Rightarrow \varphi$ is not Σ_1^0. Likewise, consistent abduction cannot be Π_1^0, because of another reduction: this time from the validity of any formula B to True, True $\Rightarrow B$.

Consistent abduction is not Δ_2^0-complete. Although it is in Δ_2^0 and is not Π_1^0, the latter is not sufficient to prove its hardness and thereby completeness, for that we would have to show that every Δ_2^0 predicate may be reduced to consistent abduction, and we can only prove that it can be written as a conjunction of Σ_1^0 and Π_1^0, showing that it belongs to a relatively simple part of Δ_2^0.

By similar arguments we can show that the earlier weak explanatory abduction is in Δ_2^0 – and the same holds for other variants that we considered. Therefore, our strategy in this chapter of adding amendments to classical consequence is costly, as it increases its complexity. On the other hand, we seem to pay the price just once. It makes no difference with respect to complexity whether we add one or all of the abductive requirements at once. We do not have similar results about the cases with minimality and preference, as their complexity will depend on the complexity of our (unspecified) preference order.

Complexity may be lower in a number of practically important cases. First, consider poorer languages. In particular, for *propositional* logic, all our notions of abduction remain obviously decidable. Nevertheless, their fine-structure will be different. Propositional satisfiability is NP-complete, while validity is Co-NP-complete.

Another direction would restrict attention to useful fragments of predicate logic. For example, universal clauses without function symbols have a decidable consequence problem. Therefore we have the following:

PROPOSITION 4 *All our notions of abductive explanatory inference are decidable over universal clauses.*

Finally, complexity as measured in the above sense may miss out on some good features of abductive reasoning, such as possible natural bounds on search space for abducibles. A very detailed study on the complexity of logic-based abduction which takes into account different kinds of theories (propositional, clausal, Horn) as well as several minimality measures are found in [EG95].

The Role of Language

Our notions of abduction all work for arbitrary formulas, and hence they have no bias toward any special formal language. But in practice, we can often do with simpler forms. E.g., observations φ will often be atoms, and the same holds for explanations α. Here are a few observations showing what may happen.

Syntactic restrictions may make for 'special effects'. For instance, our discussion of minimal abduction contained 'Carnap's trick', which shows that the choice of $\alpha = \Theta \rightarrow \varphi$ will always do for a minimal solution. But notice that this trivialization no longer works when only atomic explanations are allowed.

Here is another example. Let Θ consist of propositional Horn clauses only. In that case, we can determine the minimal abduction for an atomic conclusion directly. A simple example will demonstrate the general method:

Let $\Theta = \{q \wedge r \rightarrow s, p \wedge s \rightarrow q, p \wedge t \rightarrow q\}$ and $\varphi = \{q\}$

$$q \wedge r \rightarrow s, p \wedge s \rightarrow q, p \wedge t \rightarrow q, \alpha? \Rightarrow q$$

(i) $\Theta, \alpha \models ((p \wedge s \rightarrow q) \wedge (p \wedge t \rightarrow q)) \rightarrow q$
(ii) $\Theta, \alpha \models (p \wedge s) \vee (p \wedge t) \vee q$

That is, first make the conjunction of all formulas in Θ having q for head and construct the implication to q (i), obtaining a formula which is already an abductive solution (a slightly simpler form than $\Theta \rightarrow \varphi$).Then construct an equivalent simpler formula (ii) of which each disjunct is also an abductive solution. (Note that one of them is the trivial one). Thus, it is relatively easier to perform this process over a simple theory rather than having to engage in a complicated reasoning process to produce abductive explanations.

Finally, we mention another partly linguistic, partly ontological issue that comes up naturally in abduction. As philosophers of science have observed, there seems to be a natural distinction between 'individual facts' and 'general laws' in explanation. Roughly speaking, the latter belong to the theory Θ, while the former occur as explananda and explanantia. But intuitively, the logical basis for this distinction does not seem to lie in syntax, but rather in the nature of things. How could we make such a distinction? ([Fla95] mentions this issue as one of the major open questions in understanding abduction, and even its implementations.) Here is what we think has to be the way to go. Explanations are sought in some specific situation, where we can check specific facts. Moreover, we adduce general laws, not tied to this situation, which involve general reasoning about the kind of situation that we are in. The latter picture is not what is given to us by classical logic. We would rather have to think of a mixed situation (as in, say, the computer program Tarski's World, cf. [BE93]), where we have two sources of information. One is direct querying of the current situation, the other general deduction (provided that it is sound with respect to this situation.) The proper format for abduction then becomes a mixture of 'theorem proving' and 'model checking' (cf. [SUM96]). Unfortunately, this would go far beyond the bounds of this book.

5. Discussion and Conclusions

Is Abductive Explanatory Inference a Logical System?

The notion of abduction as a logical inference goes back to Peirce's distinction into kinds of logical reasoning, in which abduction plays the role of hypothetical inference. Therefore, its certainty is low and its non-monotonicity high. Though these aspects make it difficult to be handled, it is certainly a logical system of its own kind, which may be classified of the inductive type within Haack's approach[9]. It shares the language with classical logic and abductive conclusions are not valid by only means of the classical consequence (but abduction may have the underlying consequence relation of the deductive type). It can even be somewhat identified with the second characterization of induction, namely in which it is improbable 'supposing'/given that the 'hypotheses'/premisses are true that the conclusion is false, and therefore we can assert that the conclusion is true in a tentative way. But abduction may also be classified as a deviant system, such as in the explanatory abductive version, in which the premisses really contribute for asserting the conclusion, they are relevant in a way to the conclusion. In other words, the language is the same, but the consequence relation is more demanding to the conclusion.

The various types of abductive explanatory styles in a larger universe of other deductive and inductive systems of logic naturally commits us to a pluralistic and a global view of logic, such as Haack's own position, in which there is a variety of logical systems which rather than competing and being rival to each other, they are complementary in that each of them has a specific notion of validity corresponding to an extra-systematic one and a rigorous way for validating arguments, for it makes sense to speak of a logical system as correct or incorrect, having several of them. And finally, the global view states for abduction that it must aspire to global application, irrespective of subject-matter, and thus found in scientific reasoning and in common sense reasoning alike.

Abductive Explanatory Inference as a Structured Logical Inference

Studying abduction as a kind of logical inference has provided much more detail to the broad schema in the previous chapters. Different conditions for a formula to count as a genuine explanation, gave rise to different abductive styles of inference. Moreover, the latter can be used over different underlying

[9] Although Haack does not include explicitly abduction in her classification, she admits her existence [Haa78, page 12n].

notions of consequence (classical, preferential, statistical). The resulting abductive explanatory logics have links with existing proposals in the philosophy of science, and even further back in time, with Bolzano's notion of deducibility. They tend to be non-monotonic in nature. Further logical analysis of some key examples revealed many further structural rules. In particular, consistent abduction was completely characterized. Finally, we have discussed possible complete systems for special kinds of abduction, as well as the complexity of abduction in general.

The analysis of abductive explanatory inference at such an abstract meta-logical level, has allowed for an outlook from a purely structural perspective. We have taken its bare bones and study its consequence type with respect to itself (reflexivity), to the ability to handle new information (monotonicity), to the loss of repeated information (contraction), to the order in which premisses appear (permutation) and to the ability of handling chains of arguments (cut). In short, we tested abductive inference with respect to its ability to react to a changing world. As it turned out, none of the above properties were really an issue for plain abduction, for it is ruled by classical consequence, and therefore observes the same behaviour as that of classical reasoning. It easily allows reflexivity and new information does not invalidate in any way previous one. Moreover, in plain abduction premisse order does not affect the outcome of reasoning. Finally, cut is an easy rule to follow. In contrast, consistent abduction is not even classically reflexive, Still, every formula that in turn explains consistently (or it is explained consistently by something else) is reflexive, and thus ensuring that consistency is preserved. Consistent abduction is also very sensitive to the growth of information, as inconsistent information cannot come in at all. But if the new data explains something else together with the theory, then it is possible to add it as new information. Finally, consistent abduction also handles a somewhat sophisticated kind of cut, giving thus a way to chain the arguments. The only rules, which these two types of abductive reasoning share, are contraction of repeated formulae and permutation.

Here is what we consider the main outcomes of our analysis. We can see abductive explanatory inference as a more structured form of consequence, whose behavior is different from classical logic, but which still has clear inferential structure. The modifications of classical structural rules, which arise in this process, may even be of interest by themselves – and we see this whole area as a new challenge to logicians. Note that we did not locate the 'logical' character of abduction in any specific set of (modified) structural rules. If pressed, we would say that some modified versions of Reflexivity, Monotonicity and Cut seem essential – but we have not been able to find a single formulation that would stand once and for all. (Cf. [Gab94a] and [Gab94b] for a fuller discussion of the latter point.) Another noteworthy point was our ternary format of inference, which gives different roles to the theory and explanation on the one hand, and to the conclusion on the other. This leads to finer-grained views of inference rules, whose interest has been demonstrated.

Abductive Explanatory Inference and Geometries

Nevertheless, the structural characterization we have proposed still leaves a question unanswered, namely, in what sense an structural characterization leads to a logic, to a full syntactic or semantic characterization. Even more, despite the technical results presented, some readers may still doubt whether abductive reasoning can be considered really *logical*, perhaps it is more appropriate to render it as a special type of reasoning. After all, by accepting abductive reasoning as logical we are accepting a system that only produces tentative conclusions and not certainties as it is the case for classical reasoning. Let me precise these questions as follows:

1 In what sense the structural characterization of consistent abduction does lead to its logic?

2 Are non-classical inferences, such as abduction, really *logical*?

Regarding the first of these questions, its answer concerns a mathematical technical problem. That is, it implies a reformulation of the representation theorem into a completeness theorem, for a logical language without operators (recall that structural rules are pure, they have no connectives). Furthermore, a syntactic characterization of abduction requires the extension of the logical language, to include axioms and operators in order to formulate rules with connectives and so construct an adequate logical abductive calculi. This way to proceed, which is to obtain a syntax out of an structural characterization, has been explored with success for other logics, such as dynamic, relevance and categorial. Regarding a semantics for abduction, there is also some exploratory work in this direction, using an extended version of semantic tableaux (cf. next chapter). However, I conjecture that an abductive version such as the one allowing all conditions at once does produce a logic that is incomplete.

Regarding the second question, its answer concerns a terminological question of what we want to denote by the term *logic*. Although structural analysis of consequence has proved very fruitful and has even been proposed as a distinguished enterprise of *Descriptive Logic* in [Fla95], many logicians remain doubtful, and withhold the status of bona fide 'logical inference' to the products of non-standard styles.

This situation is somewhat reminiscent of the emergence of non-euclidean geometries in the nineteenth century. Euclidean geometry was thought of as the one and only geometry until the fifth postulate (the parallel axiom) was rejected, giving rise to new geometries. Most prominently, the one by Lobachevsky , which admits of more than one parallel, and the one by Riemann admitting none. The legitimacy of these geometries was initially doubted but their impact gradually emerged[10]. In our context, it is not geometry but styles of reasoning that occupy the space, and there is not one postulate under critical scrutiny, but several. Rejecting monotonicity gives rise to the family of non-monotonic logics, and rejecting permutation leads to styles of dynamic inference. Linear logics on the other hand, are created by rejecting contraction. All these alternative logics might get their empirical vindication, too – as reflecting different *modes* of human reasoning.

Whether non-classical modes of reasoning are really logical is like asking if non-euclidean geometries are really geometries. The issue is largely terminological, and we might decide – as Quine did on another occasion (cf.[Qui61]) – to *just* give conservatives the word 'logic' for the more narrowly described variety, using the word 'reasoning' or some other suitable substitute for the wider brands. In any case, an analysis in terms of structural rules does help us to bring to light interesting features of abduction, logical or not.

Conclusions

Summarizing, we have shown that abduction can be studied with profit as a purely logical notion of inference. Of course, we have not exhausted this viewpoint here – but we must leave its full exploration to other logicians. Also, we do not claim that this analysis exhausts all essential features of abduction, as discussed in chapter 2. To the contrary, there are clear limitations to what our present perspective can achieve. While we were successful in characterizing what an explanation is, and even show how it should behave inferentially under addition or deletion of information, the generation of abductive explanations was not discussed at all. The latter procedural enterprise is the topic of our next chapter. Another clear limitation is our restriction to the case of 'novelty', where there is no conflict between the theory and the observation. For the case of 'anomaly', we need to go into theory revision, as will happen in chapter 8. That chapter will also resume some threads from the present one, including a full version of abduction, in which all our cumulative conditions are incorporated. The latter will be needed for our discussion of Hempel's deductive-nomological model of explanation.

Related Work

Abduction has been recognized as a non-monotonic logic but with few exceptions, no study has been made to characterize it as a logical inference. In [Kon90] a general theory of abduction is defined as classical inference with

[10] The analogy with logic can be carried even further, as these new geometries were sometimes labeled 'meta-geometries'.

the additional conditions of consistency and minimality, and it is proved to be implied by Reiter's causal theories [Rei87], in which a diagnosis is a minimal set of abnormalities that is consistent with the observed behaviour of a system. Abduction is also proposed as a procedural mechanism in which the input Q is an abductive stimulus (goal), and we are interested in Δ' such that $\Delta + \Delta'$ explains Q (with some suitable underlying inference) [Gab94b, p. 199].

Another approach, closer to our own, though developed independently, is found in Peter Flach's PhD dissertation "Conjectures: an inquiry concerning the logic of induction" [Fla95], which we will now briefly describe and compare to our work (some of what follows is based on a more recent version of his proposal [Fla96a].)

Flach's logic of induction

Flach's thesis is concerned with a logical study of conjectural reasoning, complemented with an application to relational databases. An inductive consequence relation $\prec$ ($\prec \subseteq LxL$, L a propositional language) is a set of formulae; $\alpha \prec \beta$ interpreted as "β is a possible inductive hypothesis that *explains* α", or as: "β is a possible inductive hypothesis *confirmed by* α". The main reason for this distinction is to dissolve the paradoxical situation posed by Hempel's adequacy conditions for confirmatory reasoning [Hem43, Hem45], namely that in which a piece of evidence E could confirm any hypothesis whatsoever[11]. Therefore, two systems are proposed: one for the logic of confirmation and the other for the logic of explanation, each one provided with an appropriate representation theorem for its characterization. These two systems share a set of inductive principles and differ mainly in that explanations may be strengthened without ceasing to be explanations (H5), and confirmed hypotheses may be weakened without being disconfirmed (H2). To give an idea of the kind of principles these systems share, we show two of them, the well-known principles of verification and falsification in Philosophy of Science:

I1 If $\alpha \prec \beta$ and $\models \alpha \wedge \beta \rightarrow \gamma$, then $\alpha \wedge \gamma \prec \beta$.

I2 If $\alpha \prec \beta$ and $\models \alpha \wedge \beta \rightarrow \gamma$, then $\alpha \wedge \neg\gamma \not\prec \beta$.

They state that when a hypothesis β is tentatively concluded on the basis of evidence α, and a prediction γ drawn from α and β is observed, then β counts as a hypothesis for both α and γ (I1), and not for α and $\neg\gamma$ (I2) (a consequence of the latter is that reflexivity is only valid for consistent formulae).

[11]This situation arises from accepting reflexivity (H1: any observation report is confirmed by itself) and stating on the one hand that if an observation report confirms a hypothesis, then it also confirms every consequence of it (H2), and on the other that if an observation report confirms a hypothesis, then it also confirms every formula logically entailing it (H5).

Comparison to our work

Despite differences in notation and terminology, Flach's approach is connected to ours in several ways. Its philosophical motivation is based on Peirce and Hempel, its methodology is also based on structural rules, and we agree that the relationship between explananda and explanandum is a logical parameter (rather than fixed to deduction) and on the need for complementing the logical approach with a computational perspective. Once we get into the details however, our proposals present some fundamental differences, from a philosophical as well as a logical point of view.

Flach departs from Hempel's work on confirmation [Hem43, Hem45], while ours is based on later proposals on explanation [HO48, Hem65]. This leads to a discrepancy in our basic principles. One example is (consistent) reflexivity; a general inductive principle for Flach but rejected by us for explanatory abduction (since one of Hempel's explanatory adequacy conditions imply that it is invalid, cf. chapter 5). Note that this property reflects a more fundamental difference between confirmation and explanation than H2 and H5: evidence confirms itself, but it does not explain itself [12]. There are also differences in the technical setup of our systems. Although Flach's notion of inductive reasoning may be viewed as a strengthened form of logical entailment, the representation of the additional conditions is explicit in the rules rather than within the consequence relation. Nevertheless, there are interesting analogies between the two approaches, which we must leave to future work. We conclude with a general remark. A salient point in both our approaches is the importance of consistency, also crucial in Hempel's adequacy conditions both for confirmation and explanation, and in AI approaches to abduction. Thus, Bolzano's notion of deducibility comes back as capturing an intrinsic property of conjectural reasoning in general.

[12] Flach correctly points out that Hempel's own solution to the paradox was to drop condition (H5) from his logic of confirmation. Our observation is that the fact that Hempel later developed an independent account for the logic of explanation [HO48, Hem65], suggests he clearly separated confirmation from explanation. In fact his logic for the latter differs in more principles than the ones mentioned above.

Chapter 4

ABDUCTION AS COMPUTATION

1. Introduction

Our logical analysis of abduction in the previous chapter is in a sense, purely structural. It was possible to state how abductive explanatory logic behaves, but not how abductive explanations are generated. In this chapter we turn to the question of abduction as a computational process. There are several frameworks for computing abductions; two of which are logic programming and semantic tableaux. The former is a popular one, and it has opened a whole field of abductive logic programming [KKT95] and [FK00]. The latter has also been proposed for handling abduction [MP93] and [AN04], and it is our preference here. Semantic tableaux are a well-motivated standard logical framework. But over these structures, different search strategies can compute several versions of abduction with the non-standard behaviour that we observed in the preceding chapter. Moreover, we can naturally compute various kinds of abducibles: atoms, conjunctions or even conditionals. This goes beyond the framework of abductive logic programming, in which abducibles are atoms from a special set of abducibles.

This chapter is naturally divided into six parts. After this introduction, in which we give an overview of abduction as computation, in the second part (section 2) we review the framework of semantic tableaux and propose a characterization of tableaux extensions into open, closed and semiclosed in order to prepare the ground for the construction of abductive explanations. In the third part (section 3) we first propose abduction as a process of tableau extension, with each abductive version corresponding to some appropriate 'tableau extension' for the background theory. We then translate our abductive formulations to the setting of semantic tableaux and propose two strategies for the generation of abductions. In the fourth part (section 4), we put forward algorithms to compute these different abductive versions. In particular, explanations with complex forms are constructed from simpler ones. This allows us to identify cases without consistent atomic explanations whatsoever. It also suggests that in practical implementations of abduction, one can implement our views on different abductive *outcomes* in chapter 2. The fifth part (section 5) discusses further logical aspects of tableau abduction. We analyze abductive semantic tableaux in first order logic, in particular for the case in which the corresponding formulae (corresponding to the theory and the negation of the observation) are finitely satisfiable but may generate infinite tableaux. Moreover, we present a further semantic analysis, validity of structural rules as studied in chapter 3, plus soundness and completeness of our algorithms. In the sixth and final part of this chapter (section 6), we offer an analysis of previous sections with respect to the scope and limitations of the tableaux treatment of abduction. We then put forward our conclusions and present related work within the study of abduction in semantic tableaux.

Generally speaking, this chapter shows how to implement abduction, how to provide procedural counterparts to the abductive versions described in chapter 3. There are still further uses, though which go beyond our analysis so far. Abduction as revision can also be implemented in semantic tableaux. Upcoming chapters will demonstrate this, when elaborating a connection with empirical progress in science (chapter 6) and theories of belief change in AI (chapter 8). Our formal specification can lead to the development of algorithms which may be further implemented as we have done elsewhere. (Cf. [Ali97] Appendix A, for a Prolog code).

Procedural Abduction

Computational Perspectives

There are several options for treating abduction from a procedural perspective. One is standard proof analysis, as in logical proof theory (cf. [Tro96]) or in special logical systems that depend very much on proof-theoretic motivations, such as relevance logic, or linear logic. Proof search via the available rules would then be the driving force for finding abducibles. Another approach would program purely computational algorithms to produce the various types of abduction that we want. An intermediate possibility is logic programming, which combines proof theory with an algorithmic flavor. The latter is more in line with our view of abductive logic as inference plus a control strategy (cf. chapter 2). Although we will eventually choose yet a different route toward the latter end, we do sketch a few features of this practically important approach.

Abducing in Logic Programming

Computation of abductions in logic programming can be formulated as the following process. We wish to produce literals $\alpha_1, \ldots, \alpha_n$ which, when added to the current program P as new facts, make an earlier failed goal φ (the 'surprising fact') succeed after all via Prolog computation $\Rightarrow_p$:

α is an abductive explanation for query φ given program P if:

$P \not\Rightarrow_p \varphi$ while $\alpha, P \Rightarrow_p \varphi$

Notice that we insert the abducibles as facts into the program here - as an aid. It is a feature of the Prolog proof search mechanism, however, that other positions might give different derivational effects. In this chapter, we merely state these, and other features of resolution-style theorem proving without further explanation, as our main concerns lie elsewhere. (Cf. [FK00] for several papers on abductive logic programming).

Two Abductive Computations

Abductions are produced via the PROLOG resolution mechanism and then checked against a set of 'potential abducibles'. But as we just noted, there are several ways to characterize an abductive computation, and several ways to add a fact to a Prolog program. An approach which distinguishes between two basic abductive procedures is found in [Sti91], who defines *most specific abduction* (MSA) and *least specific abduction* (LSA). These differ as follows. Via MSA only pure literals are produced as abductions, and via LSA those that are not. A 'pure literal' is one which may occur both in the body of a clause as well as in the head, whereas a 'non-pure literal' for a program P, is one which cannot be resolved via any clause in the program and thus may appear only in the body of a clause. The following example illustrates these two procedures (it is a combination of our earlier common sense rain examples):

Program P**:** $r \leftarrow c, w \leftarrow r, w \leftarrow s$
Query q: w
MSA: c, s
LSA: r

This distinction is useful when we want to identify those abductions that are 'final causes' (MSA) from 'indirect causes', which may be explained by something else (LSA).

Structural Rules

This type of framework also lends itself to a study of logical structural rules like in chapter 3. This time, non-standard effects may reflect computational peculiarities of our proof search procedure. (Cf. [Min90, Kal95, vBe92, Gab94b] for more on this general phenomenon.) As for Monotonicity, we have already shown (cf. chapter 3) that right- but not left-insertion of new clauses in a program is valid for Prolog computation. Thus, adding an arbitrary formula at the beginning of a program may invalidate earlier programs. (With inserting atoms, we can be more liberal: but cf. [Kal95] for pitfalls even there.) For another important structural rule, consider Reflexivity. It is valid in the following form:

Reflexivity: $\alpha, P \Rightarrow_p \alpha$

but invalid in the form

Reflexivity: $P, \alpha \Rightarrow_p \alpha$

Moreover, these outcomes reflect the downward computation rule of Prolog. Other algorithms can have different structural rules[1].

2. Semantic Tableaux

The Classical Method

Tableau Construction

The logical framework of semantic tableaux is a refutation method introduced in the 50's independently by Beth [Bet69] and Hintikka [Hin55]. A more modern version is found in [Smu68] and it is the one presented here, although it will be referred to as 'Beth's Tableaux'. The general idea of semantic tableaux is as follows:

> To test if a formula φ follows from a set of premises Θ, a *tableau tree* for the sentences in $\Theta \cup \{\neg\varphi\}$ is constructed, denoted by $\mathcal{T}(\Theta \cup \{\neg\varphi\})$. The tableau itself is a binary tree built from its initial set of sentences by using rules for each of the logical connectives that specify how the tree branches. If the tableau closes (every branch contains an atomic formula ψ and its negation), the initial set is unsatisfiable and the entailment $\Theta \models \varphi$ holds. Otherwise, if the resulting tableau has open branches, the formula φ is not a valid consequence of Θ.

The rules for constructing the tableau tree are as follows. There are seven rules, which may be reduced to two general types of transformation rules, one 'conjunctive' (α-type) and one 'disjunctive' (β-type). The former for a true conjunction and the latter for a true disjunction suffice if every formula to be incorporated into the tableau is transformed first into a propositional conjunctive or a disjunctive normal form. Moreover, the effect of applying an α-type rule is that of adding α_1 followed by α_2 in the open branches, while the resulting operation of an β-type rule is to generate two branches, β_1 and β_2 in each of the open branches. Our notation is as follows:

[1] In particular, success or failure of Reflexivity may depend on whether a 'loop clause' $\alpha \leftarrow \alpha$ is present in the program. Also, Reflexivity is valid under MSA, but not via LSA computation, since a literal α in a rule $\alpha \leftarrow \alpha$ is a pure literal.

Rule A:

$$\alpha \quad \longrightarrow \quad \frac{\alpha_1}{\alpha_2}$$

Rule B:

$$\beta \quad \longrightarrow \quad \beta_1 \mid \beta_2$$

The seven rules for tableaux construction are the following. Double negations are suppressed. True conjunctions add both conjuncts, negated conjunctions branch into two negated conjuncts. True disjunctions branch into two true disjuncts, while negated disjunctions add both negated disjuncts. Implications $(a \to b)$ are treated as disjunctions $(\neg a \vee b)$.

- Negation

$$\neg\neg\varphi \quad \longrightarrow \quad \varphi$$

- Conjunction

$$\varphi \wedge \psi \quad \longrightarrow \quad \frac{\varphi}{\psi}$$

$$\neg(\varphi \wedge \psi) \quad \longrightarrow \quad \neg\varphi \mid \neg\psi$$

- Disjunction

$$\varphi \vee \psi \quad \longrightarrow \quad \varphi \mid \psi$$

$$\neg(\varphi \vee \psi) \quad \longrightarrow \quad \frac{\neg\varphi}{\neg\psi}$$

- Implication

$$\varphi \to \psi \quad \longrightarrow \quad \neg\varphi \mid \psi$$

$$\neg(\varphi \to \psi) \quad \longrightarrow \quad \frac{\varphi}{\neg\psi}$$

A Simple Example

To show how this works, we give an extremely simple example. More elaborate tableaux will be found in the course of this chapter. Let $\Theta = \{r \to w\}$. Set $\varphi = w$. We ask whether $\Theta \models \varphi$. The tableau is as follows. Here, an empty circle ○ indicates that a branch is open, and a crossed circle ⊗ that the branch is closed. The tableau $\mathcal{T}(\Theta \cup \{\neg\varphi\})$ is as follows:

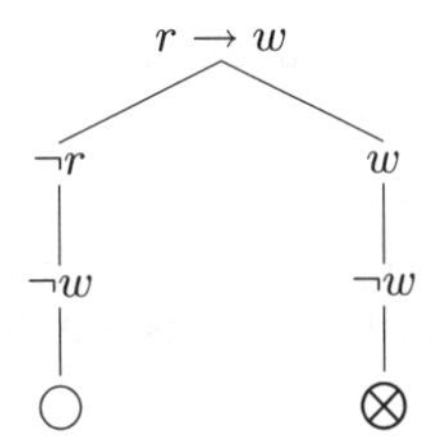

The resulting tableau is open, showing that $\Theta \not\models \varphi$. The open branch indicates a 'counterexample', that is, a case in which Θ is true while φ is false (r, w false). More generally, the construction principle is this. A tableau has to be expanded by applying the construction rules until formulas in the nodes have no connectives, and have become literals (atoms or their negations). Moreover, this construction process ensures that each node of the tableau can only carry a subformula of Θ or $\neg\varphi$.

Logical Properties

The tableau method as sketched so far has the following general properties. These can be established by simple analysis of the rules and their motivation, as providing an exhaustive search for a counter-example. In what follows, we concentrate on verifying the top formulas, disregarding the initial motivation of finding counterexamples to consequence problems. This presentation incurs no loss of generality. Given a tableau for a theory Θ, that is, $\mathcal{T}(\Theta)$:

- If $\mathcal{T}(\Theta)$ has open branches, Θ is consistent. Each open branch corresponds to a verifying model.
- If $\mathcal{T}(\Theta)$ has all branches closed, Θ is inconsistent.

Another, more computational feature is that, given some initial verification problem, the order of rule application in a tableau tree does not affect the result (in propositional logic). The structure of the tree may be different, but the outcome as to consistency is the same. Moreover, returning to the formulation with logical consequence problems, we have that semantic tableaux are a *sound and complete* system:

$$\Theta \models \varphi \quad \text{iff} \quad \text{there is a closed tableau for } \Theta \cup \{\neg\varphi\}.$$

Given the effects of the above rules, which decrease complexity, tableaux are a decision method for propositional logic. This is different with predicate logic (partially treated further in the chapter, section 5), where quantifier rules may lead to unbounded repetitions. In the latter case, the tableau method is only semi–decidable. For the propositional case we have: if the initial set of formulas is unsatisfiable, the tableau will close in finitely many steps. But if it is satisfiable, the tableau may become infinite, without terminating, recording an infinite model. In this section, we shall only consider the propositional case.

A more combinatorial observation is that there are two faces of tableaux. When an entailment does not hold, read upside down, open branches are records of counter-examples. When the entailment does hold, read bottom up, a closed tableau is easily reconstructed as a Gentzen sequent calculus proof. This is no accident of the method. In fact, Beth's tableaux may be seen as a modern formalization of the Greek methods of *analysis* and proof *synthesis*, as we already observed in chapter 1.

Tableaux are widely used in logic, and they have many further interesting properties. For a more detailed presentation, the reader is invited to consult [Smu68, Fit90]. For convenience in what follows, we give a quick reference list of some major notions concerning tableaus.

Closed Branch : A branch of a tableau is closed if it contains some formula and its negation.

Atomically Closed Branch : A branch is atomically closed if it contains an atomic formula and its negation.

Open Branch : A branch of a tableau is open if it is not closed.

Complete Branch : A branch Γ of a tableau is complete if (referring to the earlier-mentioned two main formula types) for every α which occurs in Γ, both α_1 and α_2 occur in Γ, and for every β which occurs in Γ, at least one of β_1, β_2 occurs in Γ.

Closed Tableau : A closed tableau is one in which all branches are closed.

Open Tableau : An open tableau is one in which there is at least one open branch.

Completed Tableau : A tableau $\mathcal{T}$ is completed if every branch of $\mathcal{T}$ is either closed or complete.

Proof of X : A proof of a formula X is a closed tableau for $\neg X$.

Proof of $\Theta \models \varphi$: A proof of $\Theta \models \varphi$ is a closed tableau for $\Theta \cup \{\neg\varphi\}$.

Our Extended Method

Tableaux Extensions and Closures

Here we present our proposal to extend the framework of semantic tableaux in order to compute abductions. We describe a way to represent a tableau for a theory as the union of its branches, and operations to extend and characterize tableaux extension types. A propositional language is assumed with the usual connectives, whose formulas are of three types: literals (atoms or their negations), α-type (conjunctive form), or β-type (disjunctive form).

Tableaux Representation

Given a theory Θ we represent its corresponding completed tableau $\mathcal{T}(\Theta)$ by the union of its branches. Each branch is identified with the set of formulas that label that branch. That is:

$\mathcal{T}(\Theta) = [\Gamma_1] \cup \ldots \cup [\Gamma_k]$ where each Γ_i may be open or atomically closed.

Our treatment of tableaux will be always on completed tableau, so we just refer to them as tableaux from now on.

Tableau and Branch Extension

A tableau is extended with a formula via the usual expansion rules (explained in previous subsection). An extension may modify a tableau in several ways. These depend both on the form of the formula to be added and on the other formulas in the theory represented in the original tableau. If an atomic formula is added, the extended tableau is just like the original with this formula appended at the bottom of its open branches. If the formula has a more complex form, the extended tableau may look quite different (e.g., disjunctions cause every open branch to split into two). In total, however, when expanding a tableau with a formula, the effect on the open branches can only be of three types. Either (i) the added formula closes no open branch or (ii) it closes all open branches, or (iii) it may close some open branches while leaving others open. In order to compute consistent and explanatory abductions, we need to clearly distinguish these three ways of extending a tableau. We label them as *open*, *closed*, and *semi-closed* extensions, respectively. In what follows we define these notions more precisely.

Branch Extension

Given a $\mathcal{T}(\Theta)$ the addition of a formula $\{\gamma\}$ to each of its branches Γ_i is defined by the following $+$ operation:

- $\Gamma_i + \{\gamma\} = \Gamma_i$ iff Γ_i closed:
- If Γ_i is a completed open branch:

  Case 1 $\{\gamma\}$ is a literal.
  $\Gamma_i + \{\gamma\} = \Gamma_i \cup \{\gamma\}$

  Case 2 γ is an α-type ($\gamma = \alpha_1 \wedge \alpha_2$).
  $\Gamma_i + \{\gamma\} = ((\Gamma_i \cup \{\gamma\}) + \alpha_1) + \alpha_2$

  Case 3 γ is a β-type ($\gamma = \beta_1 \vee \beta_2$).
  $\Gamma_i + \{\gamma\} = \{(\Gamma_i \cup \{\gamma\}) + \beta_1) \cup ((\Gamma_i \cup \{\gamma\}) + \beta_2\}$

That is, the addition of a formula γ to a branch is either Γ_i itself when it is closed or it is the union of its resulting branches. The operation + is defined over branches, but it easily generalizes to tableaux as follows:

- Tableau Extension:

  $\mathcal{T}(\Theta) + \{\gamma\} =_{def} \bigcup\{\Gamma_i + \{\gamma\} \mid \Gamma_i \in \mathcal{T}(\Theta)\} \qquad 1 \leq i \leq k$

Our notation allows also for embeddings $((\Theta+\gamma)+\beta)$. Note that operation $+$ is just another way of expressing the usual tableau expansion rules. Therefore, each tableau may be viewed as the result of a suitable series of $+$ extension steps, starting from the empty tableau.

Branch Extension Types

Given an open branch Γ_i and a formula γ, we have the following possibilities to extend it:

- Open Extension:

  $\Gamma_i + \{\gamma\} = \delta_1 \cup \ldots \cup \delta_n$ is open iif each δ_i is open.

- Closed Extension:

  $\Gamma_i + \{\gamma\} = \delta_1 \cup \ldots \cup \delta_n$ is closed iff each δ_i is closed.

- Semi-Closed Extension:

  $\Gamma_i + \{\gamma\} = \delta_1 \cup \ldots \cup \delta_n$ is semi-closed iff at least one δ_i is open and at least one δ_j is closed.

Extensions can also be defined over whole tableaux by generalizing the above definitions. A few examples will illustrate the different situations that may occur.

Examples

Let $\Theta = \{\neg a \vee b, c\}$.
$\mathcal{T}(\Theta) = [\neg a \vee b, \neg a, c] \cup [\neg a \vee b, b, c]$

- Open Extension: $\mathcal{T}(\Theta + \{d\})$ (d closes no branch).

  $\mathcal{T}(\Theta) + \{d\} = [\neg a \vee b, \neg a, c, d] \cup [\neg a \vee b, b, c, d]$

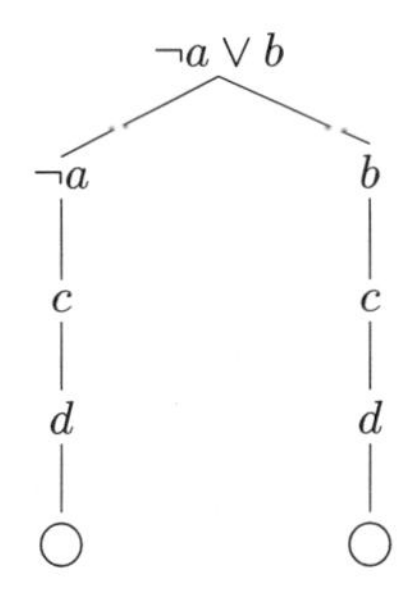

- Semi-Closed Extension: $\mathcal{T}(\Theta + \{a\})$ (a closes only one branch).

  $\mathcal{T}(\Theta) + \{a\} = [\neg a \vee b, \neg a, c, a] \cup [\neg a \vee b, b, c, a]$

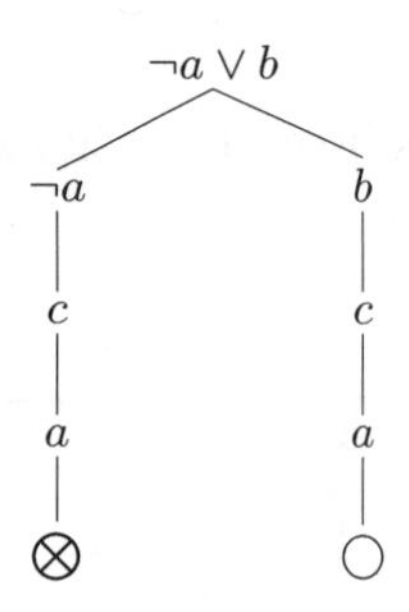

- Closed Extension: $\mathcal{T}(\Theta + \{\neg c\})$ ($\neg c$ closes all branches)

  $\mathcal{T}(\Theta) + \{\neg c\} = [\neg a \vee b, \neg a, c, \neg c] \cup [\neg a \vee b, b, c, \neg c]$

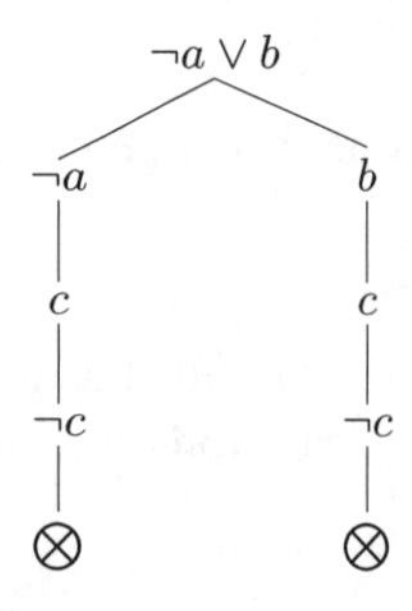

Finally, to recapitulate an earlier point, these types of extension are related to consistency in the following way:

- Consistent Extension:

  If $\mathcal{T}(\Theta) + \{\gamma\}$ is open or semi-closed, then $\mathcal{T}(\Theta) + \{\gamma\}$ is a consistent extension.

- Inconsistent Extension:

  If $\mathcal{T}(\Theta) + \{\gamma\}$ is closed, then $\mathcal{T}(\Theta) + \{\gamma\}$ is an inconsistent extension.

Given this characterization, it is easy to calculate for a given tableau, the sets of literals for each type of extension. In our example above these sets are as follows:

Open $= \{x \mid x \neq a, x \neq \neg c, x \neq \neg b\}$
Semi-closed $= \{a, \neg b\}$
Closed $= \{\neg c\}$

These constructions will be very useful for the calculations of abductions.

Branch and Tableau Closures

As we shall see when computing abductions, given a theory Θ and a formula φ, plain abductive explanations are those formulas which close the open branches of $\mathcal{T}(\Theta \cup \neg\varphi)$. In order to compute these and other abductive kinds, we need to define 'total' and 'partial closures' of a tableau. The first is the set of all literals which close every open branch of the tableau, the second of those literals which close some but not all open branches. For technical convenience, we define total and partial closures for both branches and tableaux. We also need an auxiliary notion. The negation of a literal is either its ordinary negation (if the literal is an atom), or else the underlying atom (if the literal is negative).

Given $\mathcal{T}(\Theta) = \Gamma_1 \cup \ldots \cup \Gamma_n$
(here the Γ_i are just the open branches of $\mathcal{T}(\Theta)$):

Branch Total Closure (BTC) :

The set of literals which close an open branch Γ_i:

$$\text{BTC}(\Gamma_i) = \{x \mid \neg x \in \Gamma_i\}, \qquad \text{where } x \text{ ranges over literals.}$$

Tableau Total Closure (TTC) :

The set of those literals that close all branches at once, i.e. the intersection of the BTC's:

$$TTC(\Theta) = \bigcap_{i=1}^{i=n} BTC(\Gamma_i)$$

Branch Partial Closure (BPC) :

The set of those literals which close the branch but do not close all the other open branches:

$$BPC(\Gamma_i) = BTC(\Gamma_i) - TTC(\Theta)$$

Tableau Partial Closure (TPC) :

The set formed by the union of BPC, i.e. all those literals which partially close the tableau:

$$TPC(\Theta) = \bigcup_{i=1}^{i=n} BPC(\Gamma_i)$$

In particular, the definition of BPC may look awkward, as it defines partial closure in terms of branch and tableau total closures. Its motivation lies in a way to compute *partial explanations*, being formulas which do close some branches (so they do 'explain') without closing all (so they are 'partial'). We will use the latter to construct explanations in conjunctive form.

Having defined all we need to exploit the framework of semantic tableau for our purposes, we proceed to the construction of abductive explanations.

3. Abductive Semantic Tableaux

In this section we will show the main idea for performing abduction as a kind of tableau extension. First of all, a tableau itself can represent finite theories. We show this by a somewhat more elaborate example. To simplify the notation from now on, we write $\Theta \cup \neg\varphi$ for $\Theta \cup \{\neg\varphi\}$ (that is, we omit the brackets).

Example

Let $\Theta = \{b, c \rightarrow r, r \rightarrow w, s \rightarrow w\}$, and let $\varphi = \{w\}$.
A tableau for Θ, that is $\mathcal{T}(\Theta)$, is as follows:

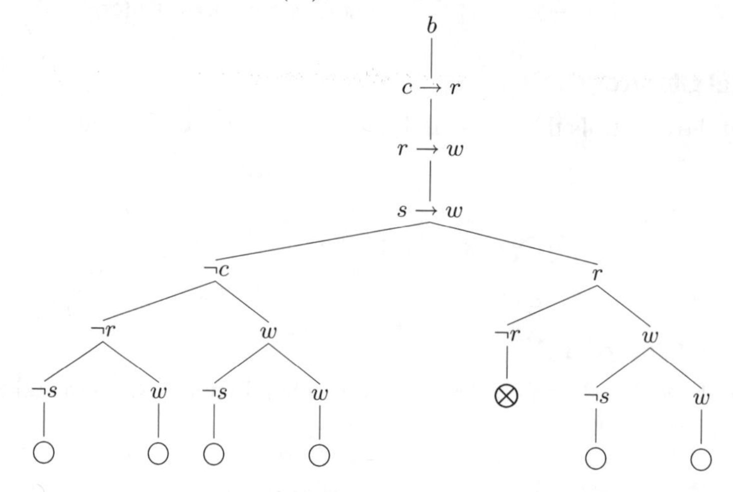

The result is an open tableau. Therefore, the theory is consistent and each open branch corresponds to a satisfying model. For example, the second branch (from left to right) indicates that a model for Θ is given by making c, r false and b, w true, so we get two possible models out of this branch (one in which s is true, the other in which it is false). Generally speaking, when constructing the tableau, the possible valuations for the formulas are depicted by the branches (either $\neg c$ or r makes the first split, then for each of these either $\neg r$ or w, and so on).

When formulas are added (thereby extending the tableau), some of these possible models may disappear, as branches start closing. For instance, when $\neg\varphi$ is added (i.e. $\neg w$), the result is the following:

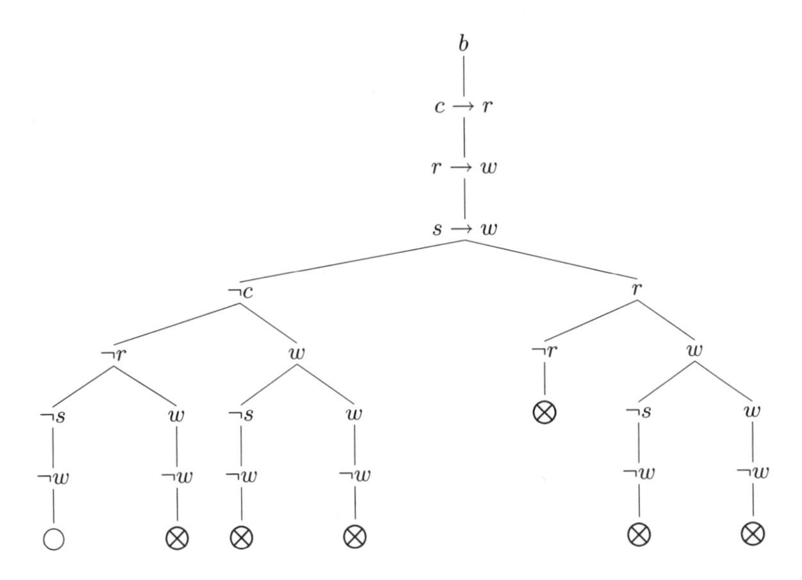

Notice that, although the resulting theory remains consistent, all but one branch has closed. In particular, most models we had before are no longer satisfying, as $\neg w$ is now true as well. There is still an open branch, indicating there is a model satisfying $\Theta \cup \neg w$ (c, r, s, w false, b true), which indicates that $\Theta \not\models w$.

Abductive Semantic Tableaux: The Main Ideas

An attractive feature of the tableau method is that when φ is not a valid consequence of Θ, we get all cases in which the consequence fails, graphically represented by the open branches (as shown above, the latter may be viewed as descriptions of models for $\Theta \cup \neg\varphi$.)

This fact suggests that if these *counterexamples* were 'corrected by amending the theory', through adding more premises, we could perhaps make φ a valid consequence of some (minimally) extended theory Θ'. This is indeed the whole issue of abduction. Accordingly, abduction may be formulated in this framework as a process of *extension*, extending a tableau with suitable formulas that close the open branches.

In our example above, the remaining open branch had the following relevant (literal) part:

The following are (some) formulas whose addition to the tableau would close this branch (and hence, the whole tableau):

$$\{\neg b, c, r, s, w, c \wedge r, r \wedge w, s \wedge w, s \wedge \neg w, c \vee w\}$$

Note that several forms of statement may count here as abductions. In particular, those in disjunctive form (e.g. $c \vee w$) create two branches, which then both close. (We will consider these various cases in detail later on.). Moreover, notice that most formulae of this set (except $\neg b$ and $s \wedge \neg w$) is a semi-closed extension of the original tableau for Θ.

The Generation of Abductions

In principle, we can compute abductions for all our earlier abductive versions (cf. chapter 3). A direct way of doing so is as follows:

> First compute abductions according to the plain version and then eliminate all those that do not comply with the various additional requirements.

This strategy first translates our abductive formulations to the setting of semantic tableaux as follows:

Given Θ (a set of formulae) and φ (a sentence), α is an abductive explanation if:

Plain :

$\mathcal{T}((\Theta \cup \neg\varphi) \cup \alpha)$ is closed. $(\Theta, \alpha \models \varphi)$.

Consistent : Plain Abduction +

$\mathcal{T}(\Theta \cup \alpha)$ is open $(\Theta \not\models \neg\alpha)$

Explanatory : Plain Abduction +

(i) $\mathcal{T}(\Theta \cup \neg\varphi)$ is open $(\Theta \not\models \varphi)$

(ii) $\mathcal{T}(\alpha \cup \neg\varphi)$ is open $(\alpha \not\models \varphi)$

In addition to the 'abductive conditions' we must state constraints over our search space for abducibles, as the set of formulas fulfilling any of the above conditions is in principle infinite. Therefore, we impose restrictions on the vocabulary as well as on the form of the abduced formulas:

- Restriction on Vocabulary

  α is in the vocabulary of the theory and the observation:

  $\alpha \in \mathrm{Voc}(\Theta \cup \{\varphi\})$.

- Restriction on Form

  The syntactic form of α is either a literal, a conjunction of literals (without repeated conjuncts), or a disjunction of literals (without repeated disjuncts).

Once it is clear what our search space for abducibles is, we continue with our discussion. Note that while computation of plain abductions involves only closed branches, the other versions use inspection of both closed and open branches. For example, an algorithm computing consistent abductions would proceed in the following two steps:

- *Generating Consistent Abductions* (First Version)

  1 Generate all plain abductions, being those formulas α for which $\mathcal{T}((\Theta \cup \neg\varphi) \cup \alpha)$ is closed.

  2 Take out all those α for which $\mathcal{T}(\Theta \cup \alpha)$ is closed.

In particular, an algorithm producing consistent abductions along these lines must produce all explanations that are inconsistent with Θ. This means many ways of closing $\mathcal{T}(\Theta)$, which will then have to be removed in step 2. This is of course wasteful. Even worse, when there are no consistent explanations (besides the trivial one), so that we would want to give up, our procedure still produces the inconsistent ones. The same point holds for our other versions of abduction.

Of course, there is a preference for procedures that generate abductions in a reasonably efficient way. We will show how to devise these, making use of the representation structure of tableaux, in a way which avoids the production of inconsistent formulae. Here is our idea.

- *Generating Consistent Abductions* (Second Version)

  1 Generate all formulas α which do not close all (but some) open branches of $\mathcal{T}(\Theta)$.

  2 Check which of the formulas α produced are such that $\mathcal{T}((\Theta \cup \neg\varphi) \cup \alpha)$ is closed.

That is, first produce formulas which extend the tableau for the background theory in a consistent way, and then check which of these are abductive explanations. As we will show later, the consistent formulae produced by the second procedure are not necessarily wasteful. They might be 'partial explanations' (an ingredient for explanations in conjunctive form), or part of explanations in disjunctive form.

In other words, consistent abductions are those formulas which *"if they had been in the theory before, they would have closed those branches that remain open after $\neg\varphi$ is incorporated into the tableau"*.

In our example above the difference between the two algorithmic versions is as follows (taking into account only the atomic formulas produced). Version 1 produces a formula ($\neg b$) which is removed for being inconsistent, and version 2 produces a consistent formula ($\neg w$) which is removed for not closing the corresponding tableau.

4. Computing Abductions with Tableaux

Our strategy for computing plain abduction in semantic tableaux will be as follows. We will be using tableaux as an ordinary consequence test, while being careful about the search space for potential abducibles. The computation is divided into different forms of explanations. Atomic explanations come first, followed by conjunctions of literals, to end with those in disjunctive form. Here we sketch the main ideas for their construction, and give an example for each kind. There are no detailed algorithms of programs here, but they may be found in [Ali97] and [VA99][2].

Plain Abduction

Atomic Plain Abduction

The idea behind the construction of atomic explanations is very simple. One just computes those atomic formulas which close every open branch of $\mathcal{T}(\Theta \cup \neg\varphi)$, corresponding precisely to its Total Tableaux Closure (TTC($\Theta \cup \neg\varphi$)). Here is an example:

Let $\Theta = \{\neg a \vee b\}$ $\qquad \varphi = b.$

[2]The implementation of abduction shows that it is not particularly hard to use it in practice (which may explain its appeal to programmers). It may be a complex notion general, but when well-delimited, it poses a feasible programming task. In [Ali97] a Prolog Code (written for Arity Prolog) may be found of the main procedures for implementing plain abduction (atomic, conjunctive and disjunctive explanations altogether). In [VA99] the essentials for an implementation in the java language is offered for our abductive strategy (the first version) given in the previous section. This turns out to be a good choice, for the high level of abstraction to describe objects and classes in object oriented programming, allows for an straightforward mapping between the formal tableaux and operations and their corresponding functions and procedures. This is an advantage over implementations in logic programming, in which the process is tied to the resolution rule.

$\mathcal{T}(\Theta \cup \neg\varphi)$ is as follows:

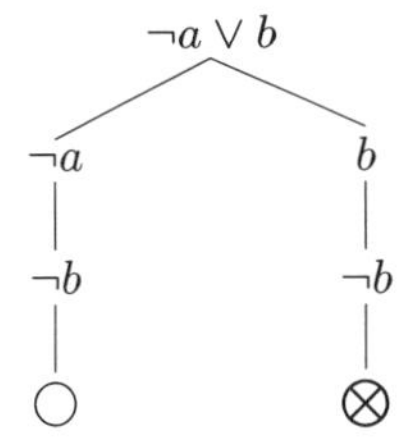

The two possible atomic plain abductions are $\{a, b\}$.

Conjunctive Plain Abduction

Single atomic explanations may not always exist, or they may not be the only ones of interest. The case of explanations in conjunctive form ($\alpha = \alpha_1 \wedge \ldots \wedge \alpha_n$) is similar to the construction of atomic explanations. We look for literals that close branches, but in this case we want to get the literals that close some but not all of the open branches. These are the conjuncts of a 'conjunctive explanation', and they belong to the tableau partial closure of Θ (i.e., to TPC($\Theta \cup \neg\varphi$)). Each of these *partial explanations* make the fact φ 'less surprising' by closing some (but not all) of the open branches. Together they constitute an abductive explanation.

As a consequence of this characterization, no partial explanation is an atomic explanation. That is, a conjunctive explanation must be a conjunction of partial explanations. The motivation is this. We want to construct explanations which are *non-redundant*, in which every literal does some explaining. Moreover, this condition allows us to bound the production of explanations in our algorithm. We do not want to create what are intuitively 'redundant' combinations. For example, if p and q are abductive explanations, then $p \wedge q$ should not be produced as explanation. Thus we impose the following condition:

Non-Redundancy

Given an abductive explanation α for a theory Θ and a formula φ, α is *non-redundant* if it is either atomic, or no subformula of α (different from φ) is an abductive explanation.

The following example gives an abductive explanation in conjunctive form which is non-redundant:

Let $\Theta = \{\neg a \vee \neg c \vee b\}$, and $\varphi = b$.

The corresponding tableau $\mathcal{T}(\Theta \cup \neg\varphi)$ is as follows:

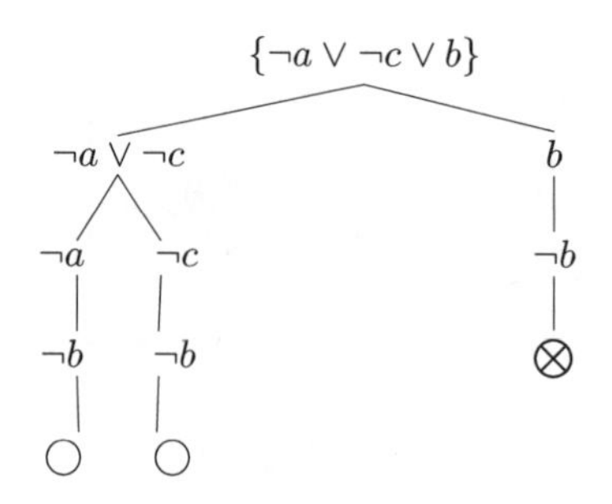

The only atomic explanation is the trivial one $\{b\}$. The conjunctive explanation is $\{a \wedge c\}$ of which neither conjunct is an atomic explanation.

Disjunctive Plain Abductions

To stay in line with computational practice, we shall sometimes regard abductive explanations in disjunctive form as implications. (This is justified by the propositional equivalence between $\neg \alpha_i \vee \alpha_j$ and $\alpha_i \rightarrow \alpha_j$.) These special explanations close a branch by splitting it first into two. Disjunctive explanations are constructed from atomic and partial explanations. We will not analyze this case in full detail, but provide an example of what happens.

Let $\Theta = \{a\}$ $\quad \varphi = b$.

The tableau structure for $\mathcal{T}(\Theta \cup \neg b)$ is as follows:

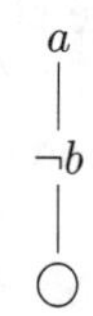

Notice first that the possible atomic explanations are $\{\neg a, b\}$ of which the first is inconsistent and the second is the trivial solution. Moreover, there are no 'partial explanations' as there is only one open branch. An explanation in disjunctive form is constructed by combining the atomic explanations, that is , $\{\neg a \vee b\}$. The effect of adding it to the tableau is as follows ($\mathcal{T}(\Theta \cup \{\neg b\} \cup \{\neg a \vee b\})$):

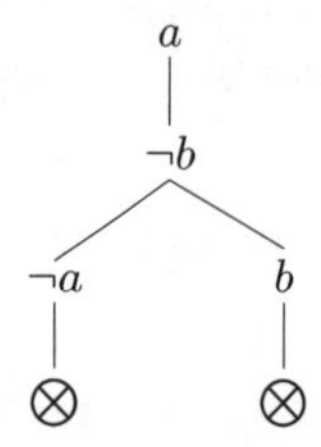

This examples serves as a representation of our example in chapter 2, in which a causal connection is found between certain types of clouds (a) and rain (b), namely that a causes b ($a \rightarrow b$).

Algorithm for Computing Plain Abductions

The general points of our algorithm for computing plain abductions is displayed here.

- **Input:**
  . A set of propositional formulas separated by commas representing the theory Θ.
  . A literal formula φ representing the 'fact to be explained'.
  . Preconditions: Θ, φ are such that $\Theta \not\models \varphi$, $\Theta \not\models \neg\varphi$.
- **Output:**
  Produces the set of abductive explanations: $\alpha_1, \ldots \alpha_n$ such that:
  (i) $\mathcal{T}((\Theta \cup \neg\varphi) \cup \alpha_i)$ is closed.
  (ii) α_i complies with the vocabulary, form restrictions as well as with the non-redundancy condition.
- **Procedure:**
  . Calculate $\Theta + \neg\varphi = \{\Gamma_1, \ldots, \Gamma_k\}$
  . Take those Γ_i which are open branches: $\Gamma_1, \ldots, \Gamma_n$
  . **Atomic Plain Explanations**

    1 Compute TTC($\Gamma_1, \ldots, \Gamma_n$)= $\{\gamma_1, \ldots, \gamma_m\}$.

    2 $\{\gamma_1, \ldots, \gamma_m\}$ is the set of atomic plain abductions.

  . **Conjunctive Plain Explanations**

    1 For each open branch Γ_i, construct its partial closure: BPC(Γ_i).

    2 Check if all branches Γ_i have a partial closure, for otherwise there cannot be a conjunctive solution (in which case, goto END).

    3 Each BPC(Γ_i) contains those literals which partially close the tableau. Conjunctive explanations are constructed by taking one literal of each BPC(Γ_i) and making their conjunction. A typical solution is a formula β as follows:
    $a_1 \wedge b_1 \wedge \ldots \wedge z_1 \qquad (a_1 \in BPC(\Gamma_1), b_1 \in BPC(\Gamma_2), \ldots, z_1 \in BPC(\Gamma_n))$

    4 Each β conjunctive solution is reduced (there may be repeated literals). The set of solutions in conjunctive form is $\beta_1, \ldots, \beta_l$[3].

    5 END.

  . **Disjunctive Plain Explanations**

    1 Construct disjunctive explanations by means of the following combinations: atomic explanations amongst themselves, conjunctive explanations amongst themselves, conjunctive with atomic, and each of atomic and conjunctive with φ. We just show two of these constructions:

[3]There are no redundant solutions. The reason is that each conjunctive solution β_i is formed by a conjunction of 'potential explanations', none of which is itself an abductive solution.

2 Generate pairs from the set of atomic explanations, and construct their disjunctions $(\gamma_i \vee \gamma_j)$.

3 For each atomic explanation γ_i construct the disjunction with φ as follows: $(\gamma_i \vee \varphi)$.

4 The result of all combinations above is the set of explanations in disjunctive form.

5 Verify the previous set for redundant explanations. Take out all of these. The remaining are the explanations in disjunctive form.

6 END.

Consistent Abduction

The issue now is to compute abductive explanations with the additional requirements of being consistent. For this purpose we will follow the same presentation as for plain abductions (atomic, conjunctive and disjunctive), and will give the key points for their construction. Our algorithm follows version 2 of the strategies sketched earlier (cf. subsection 3.2). That is, it first constructs those consistent extensions on the original tableau for Θ which do some closing and then checks which of these is in fact an explanation (i.e. closes the tableau for $\Theta \cup \neg\varphi$). This way we avoid the production of any inconsistency whatsoever. It turns out that in the atomic and conjunctive cases explanations are sometimes necessarily inconsistent, therefore we identify these cases and prevent our algorithm from doing anything at all (so that we do not produce formulae which are discarded afterwards).

Atomic Consistent Abductions

When computing plain atomic explanations, we now want to avoid any computation when there are only inconsistent atomic explanations (besides the trivial one). Here is an observation which helps us get one major problem out of the way. Atomic explanations are necessarily inconsistent when $\Theta + \neg\varphi$ is an open extension (recall that φ is a literal). So, we can prevent our algorithm from producing anything at all in this case.

FACT 1 *Whenever $\Theta + \neg\varphi$ is an open extension, and α a non-trivial atomic abductive explanation (i.e. different from φ), it follows that Θ, α is inconsistent.*

Proof. Let $\Theta + \neg\varphi$ be an open extension and α an atomic explanation ($\alpha \neq \varphi$). The latter implies that $((\Theta + \neg\varphi) + \alpha)$ is a closed extension. Therefore, $\Theta + \alpha$ must be a closed extension, too, since $\neg\varphi$ closes no branches. But then, $\Theta + \alpha$ is an inconsistent extension. I.e. Θ, α is inconsistent. $\dashv$

This result cannot be generalized to more complex forms of abducibles. (We will see later that for explanations in disjunctive form, open extensions need not lead to inconsistency.) In case $\Theta + \neg\varphi$ is a semi-closed extension, we have to do real work, however, and follow the strategy sketched above. The key point in the algorithm is this. Instead of building the tableau for $\Theta \cup \varphi$ directly, and working with its open branches, we must start with the open branches of Θ.

Algorithm

1 $\Theta, \neg\varphi$ are given as input and are such that: $\Theta \not\models \varphi, \Theta \not\models \neg\varphi$.

2 Calculate $\Theta + \neg\varphi$. If it is an open extension, then there are no atomic consistent explanations (by fact 1), go to Step 6.

3 Calculate the set of literals $\{\gamma_1, \ldots, \gamma_n\}$ for which $\Theta + \gamma_i$ is a semi-closed extension.

4 Select from above set those γ_i for which $(\Theta + \neg\varphi) + \gamma_i$ is a closed extension.

5 The set above is the set of Consistent Atomic Explanations.

6 END.

Conjunctive Consistent Explanations

For conjunctive explanations, we can also avoid any computation when there are only 'blatant inconsistencies', essentially by the same observation as before.

FACT 2 *Whenever $\Theta + \neg\varphi$ is an open extension, and $\alpha = \alpha_1 \wedge \ldots \wedge \alpha_n$ is a conjunctive abductive explanation, it holds that Θ, α is inconsistent.*

The proof is analogous to that for the atomic case.

In order to construct partial explanations, the ingredients of conjunctive explanations, the key idea is to build conjunctions out of those formulas for which both $\Theta + \gamma_i$ and $(\Theta + \neg\varphi) + \gamma_i$ are semiclosed extensions. The reason is as follows: the former condition assures that these formulas do close at least one branch from the original tableau. The latter condition discards atomic explanations (for which $(\Theta + \neg\varphi) + \gamma_i$ is closed) and the trivial solution ($\gamma_i = \varphi$ when $(\Theta + \neg\varphi) + \gamma_i$ is open), and takes care of non-redundancy as a side effect. Conjunctions are then constructed out of these formulas and those which induce a closed extension for $\Theta + \neg\varphi$ are the selected conjunctive explanations.

Here is the algorithm sketch:

1 $\Theta, \neg\varphi$ are given as input and are such that: $\Theta \not\models \varphi, \Theta \not\models \neg\varphi$.

2 Calculate $\Theta + \neg\varphi$. If it is an open extension, then there are no conjunctive consistent explanations (by fact 2), go to Step 8.

3 Calculate the set of literals $\{\gamma_1, \ldots, \gamma_n\}$ for which $\Theta + \gamma_i$ is a semi-closed extension.

4 Select from above set those γ_i for which $(\Theta + \neg\varphi) + \gamma_i$ is a semi-closed extension. This is the set of Partial Explanations: $\{\gamma_1, \ldots, \gamma_m\}$, $m \leq n$.

5 Construct conjunctions from set of partial explanations starting in length k (number of open branches) to end in lenght m (number of partial explanations). Label these conjunctions as follows: $\rho_1, \ldots, \rho_s$.

6 Select those ρ_i from above for which $(\Theta + \neg\varphi) + \rho_i$ is a closed extension.

7 The set above is the set of Consistent Conjunctive Explanations.

8 END.

What this construction suggests is that there are always consistent explanations in disjunctive form, provided that the theory is consistent:

FACT 3 *Given that $\Theta \cup \neg\varphi$ is consistent, there exists an abductive consistent explanation in disjunctive form.*

The key point to prove this fact is that an explanation may be constructed as $\alpha = \neg X \vee \varphi$, for any $X \in \Theta$.

Explanatory Abduction

As for explanatory abductions, recall these are those formulas α which are only constructed when the theory does not explain the observation already ($\Theta \not\models \varphi$) and that cannot do the explaining by themselves ($\alpha \not\models \varphi$), but do so in combination with the theory ($\Theta, \alpha \models \varphi$).

Given our previous algorithmic constructions, it turns out that the first condition is already 'built-in', since all our procedures start with the assumption that the tableau for $\Theta \cup \neg\varphi$ is open[4]. As for the second condition, its implementation actually amounts to preventing the construction of the trivial solution ($\alpha = \varphi$). Except for this solution, our algorithms never produce an α such that $\alpha \not\models \varphi$, as proved below:

FACT 4 *Given any Θ and φ, our algorithm never produces abductive explanations α with $\alpha \models \varphi$ (except for $\alpha = \varphi$).*

Proof. Recall our situation: $\Theta \not\models \varphi$, $\Theta, \alpha \models \varphi$, while we have made sure Θ and α are consistent. Now, first suppose that α is a literal. If $\alpha \models \varphi$, then $\alpha = \varphi$ which is the trivial solution. Next, suppose that α is a conjunction produced by our algorithm. If $\alpha \models \varphi$, then one of the conjuncts must be φ itself. (The only other possibility is that α is an inconsistent conjunction, but this is ruled out by our consistency test.) But then, our non-redundancy filter would have produced the relevant conjunct by itself, and then rejected it for triviality. Finally, suppose that α is a disjunctive explanation. Given the above conditions tested in our algorithm, we know that Θ is consistent with at least one disjunct α_i. But also, this disjunct by itself will suffice for deriving φ in the presence of Θ, and it will imply φ by itself. Therefore, by our redundancy test, we would have produced this disjunct by itself, rather than the more complex explanation, and we are in one of the previous cases. $\dashv$

Therefore, it is easy to modify any of the above algorithms to handle the computation of explanatory abductions. We just need to avoid the trivial solution, when $\alpha = \varphi$ and this can be done in the module for atomic explanations.

Quality of Abductions

A question to ask at this point is whether our algorithms produce intuitively good explanations for observed phenomena. One of our examples (cf. disjunctive plain abductions, section 4.1) suggested that abductive explanations for a

[4]It would have been possible to take out this condition for the earlier versions. However, note that in the case that $\Theta \cup \neg\varphi$ is closed the computation of abductions is trivialized since as the tableau is already closed, any formula counts as a plain abduction and any consistent formula as a consistent abduction.

fact and a theory with no causal knowledge (with no formulas in conditional form) must be in disjunctive form if such a fact is to be explained in a consistent way. Moreover, Fact 3 stated that consistent explanations in disjunctive form are always available, provided that the original theory is consistent. However, producing consistent explanations does not guarantee that these are good or relevant. These further properties may depend upon the nature of the theory itself and of whether it makes a good combination with the observation. If the theory is a bad theory, it will produce bad or weird explanations. The following example illustrates this point.

A Bad Theory

Let $\Theta = \{r \rightarrow w, \neg r\}$ $\qquad \varphi = w \qquad \alpha = \neg r \rightarrow w.$

The tableau $\mathcal{T}(\Theta \cup \neg\varphi \cup \alpha)$ is depicted as follows:

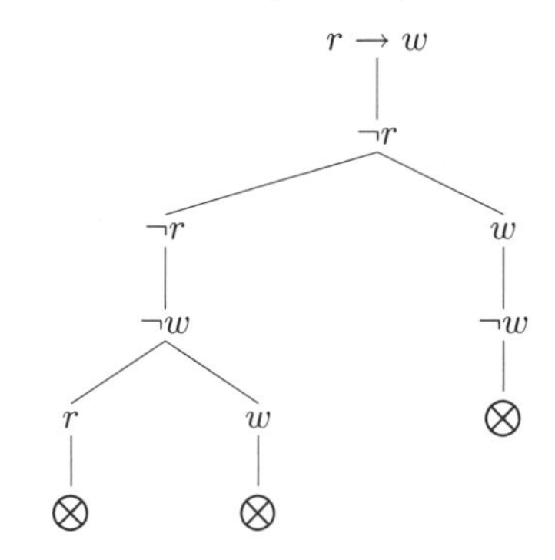

Interpreted in connection with our rain example, our algorithm will produce the following consistent 'explanation' of why the lawn is wet (w), given that rain causes the lawn to get wet ($r \rightarrow w$) and that it is not raining ($\neg r$). One explanation is that "the absence of rain causes the lawn to get wet" ($\neg r \rightarrow w$). But this explanation seems to trivialize the fact of the lawn being wet, as it seems to be so, regardless of rain!

A better, though more complex, way of explaining this fact would be to conclude that the theory is not rich enough to explain why the lawn is wet, and then look for some external facts to the theory (e.g. sprinklers are on, and they make the lawn wet.) But this would amount to dropping the vocabulary assumption.

Therefore, producing good or bad explanations is not just a business of properly defining the underlying notion of consequence, or of giving an adequate procedure. An inadequate theory like the one above can be the real cause of bad explanations. In other words, what makes a 'good explanation' is not the abducible itself, but the interplay of the abducible with the background theory and the relevant observation. Bad theories produce bad explanations. Our algorithm cannot remedy this, only show it.

Discussion

Having a module in each abductive version that first computes only atomic explanations already gives us some account of *minimal explanations* (see chapter 3), when minimality is regarded as simplicity. As for conjunctive explanations, as we have noted before, their construction is one way of computing non-trivial 'partial explanations', which make a fact less surprising by closing some, though not all open branches for its refutation. One might tie up this approach with a more general issue, namely, weaker notions of 'approximative' logical consequence. Finally, explanations in disjunctive form can be constructed in various ways. E.g., one can combine atomic explanations, or form the conjunction of all partial explanations, and then construct a conditional with φ. This reflects our view that abductive explanations are built in a compositional fashion: complex solutions are constructed from simpler ones.

Notice moreover, that we are not constructing all possible formulas which close the open branches, as we have been taking care not to produce *redundant explanations* (cf. section 4). Finally, despite these precautions, as we have noted, bad explanations may slip through when the background theory in combination with a certain observation are inappropriate.

5. Further Logical and Computational Issues

Our algorithmic tableau analysis suggests a number of further logical issues, which we briefly discuss here.

Rules, Soundness and Completeness

The abductive consequences produced by our tableaux can be viewed as a ternary notion of inference. Its structural properties can be studied in the same way as we did for the more abstract notions of chapter 3. But the earlier structural rules lose some of their point in this algorithmic setting. For instance, it follows from our tableau algorithm that consistent abduction does not allow monotonicity in either its Θ or its α argument. One substitute which we had in chapter 3 was as follows. If $\Theta, \alpha \Rightarrow \varphi$, and $\Theta, \alpha, \beta \Rightarrow \gamma$ (where γ is any conclusion at all), then $\Theta, \alpha, \beta \Rightarrow \varphi$. In our algorithm, we have to make a distinction here. We produce abducibles α, and if we already found α solving $\Theta, \alpha \Rightarrow \varphi$, then the algorithm may not produce stronger abducibles than that. (It might happen, due to the closure patterns of branches in the initial tableau, that we produce one solution implying another, but this does not have to be.) As for strengthening the theory Θ, this might result in an initial tableau with possibly fewer open branches, over which our procedure may then produce weaker abducibles, invalidating the original choice of α cooperating with Θ to derive φ.

More relevant, therefore, is the traditional question whether our algorithms are sound and complete. Again, we have to make sure what these properties mean in this setting. First, Soundness should mean that any combination $(\Theta, \alpha, \varphi)$ which gets out of the algorithm does indeed present a valid case of abduction, as defined in chapter 3. For plain abduction, it is easy to see that we have soundness, as the tableau closure condition guarantees classical consequence (which is all we need). Next, consider consistent abduction. What we need to make sure of now, is also that all abducibles are consistent with the background theory Θ. But this is what happened by our use of 'partial branch closures'. These are sure (by definition) to leave at least one branch for Θ open, and hence they are consistent with it. Finally, the conditions of the 'explanatory' algorithm ensure likewise that the theory does not explain the fact already ($\Theta \not\Rightarrow \varphi$) and that α could not do the job on its own ($\alpha \not\Rightarrow \varphi$).

Next, we consider completeness. Here, we merely make some relevant observations, demonstrating the issues (for our motives, cf. the end of this paragraph). Completeness should mean that any valid abductive consequence should actually be produced by it. This is trickier. Obviously, we can only expect completeness within the restricted language employed by our algorithm. Moreover, the algorithm 'weeds out' irrelevant conjuncts, etcetera, which cuts down outcomes even more. As a more significant source of incompleteness, however, we can look at the case of disjunctive explanations. The implications produced always involve one literal as a consequent. This is not enough for a general abductive conclusion, which might involve more. What we can say, for instance is this. By simple inspection of the algorithm, one can see that every consistent atomic explanation that exists for an abductive problem will be produced by the algorithm. In any case, we feel that completeness is less of an issue in computational approaches to abduction. What comes first is whether a given abductive procedure is natural, and simple. Whether its yield meets some pre-assigned goal is only a secondary concern in this setting.

We can also take another look at issues of soundness and completeness, relatively independently from our axiom. The following analysis of 'closure' on tableaux is inspired by our algorithm - but it produces a less procedural logical view of what is going on.

An Alternative Semantic Analysis

Our strategy for producing abductions in tableaux worked as follows. One starts with a tableau for the background theory Θ (i), then adds the negation $\neg\varphi$ of the new observation φ to its open branches (ii), and one also closes the remaining open branches (iii), subject to certain constraints. In particular (for the explanatory version) one does not allow φ itself as a closure atom as it is regarded as a trivial solution. This operation may be expressed via a kind of 'closure operation':

CLOSE $(\Theta + \neg\varphi) - \varphi$.

We now want to take an independent (in a sense, 'tableau–free') look at the situation, in the most general case, allowing disjunctive explanations. (If Θ is particularly simple, we can make do (as shown before) with atoms, or their conjunctions.) First, we recall that tableaus may be represented as sets of open branches. We may assume that all branches are completed, and hence all relevant information resides in their literals. This leads to the following observation.

FACT 5 *Let Θ be a set of sentences, and let $\mathcal{T}$ be a complete tableau for Θ, with π running over its open branches. Then $\wedge\Theta$ (the conjunction of all sentences in Θ) is equivalent to:*

$$\bigvee_{\pi \text{ open in } \mathcal{T}} \bigwedge_{l \text{ a literal}} l \in \pi.$$

This fact is easy to show. The conjunctions are the total descriptions of each open branch, and the disjunction says that any model for Θ must choose one of them. This amounts to the usual Distributive Normal Form theorem for propositional logic. Now, we can give a description of our CLOSE operation in similar terms. In its most general form, our way of closing an open tableau is really defined by putting:

$$\bigvee_{S \text{ a set of literals}} \bigwedge_{\pi \text{ open in T}} \exists l \in S : l \notin \pi$$

The inner part of this says that the set of (relevant) literals S 'closes every branch'. The disjunction states the weakest combination that will still close the tableau. Now, we have a surprisingly simple connection:

FACT 6 *CLOSE (Θ) is equivalent to $\neg(\wedge\Theta)$!*

Proof. By Fact 1 plus the propositional De Morgan laws, $\neg(\wedge\Theta)$ is equivalent to $\bigvee_{\pi \text{ open in } \mathcal{T}} \bigwedge_{l \text{ a literal}} l \notin \pi$. But then, a simple argument, involving choices for each open branch, shows that the latter assertion is equivalent to CLOSE (Θ). ⊣

In the case of abduction, we proceeded as follows. There is a theory Θ, and a surprising fact q (say), which does not follow from it. The latter shows because we have an open tableau for Θ followed by $\neg q$. We close up its open branches, without using the trivial explanation q. What this involves, as said above, is a modified operation, that we can write as:

CLOSE $(\Theta) - q$

Example

Let $\Theta = \{p \rightarrow q, r \rightarrow q\}$. The tableau $\mathcal{T}(\Theta \cup \neg q) - q$ is as follows:

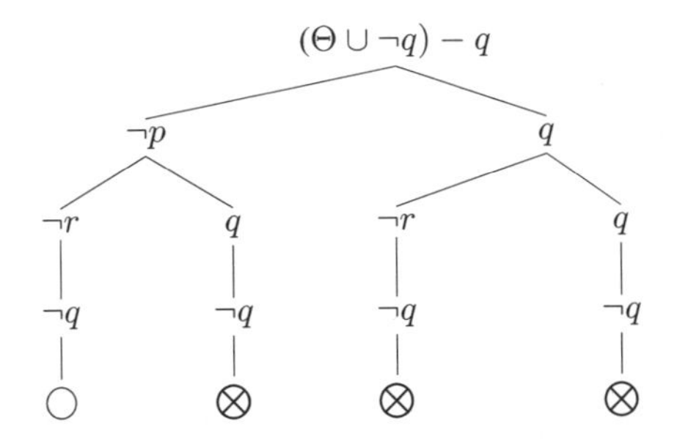

The abductions produced are p or r.

Again, we can analyze what the new operation does in more direct terms.

FACT 7 *CLOSE* $(\Theta) - q$ *is equivalent to* $\neg$ *[false/q]* $\wedge \Theta$.

Proof. From its definition, it is easy to see that CLOSE(Θ) $-$ q is equivalent with [false/q] CLOSE(Θ). But then, we have that

CLOSE $(\Theta \wedge \neg q) - q$ iff (by Fact 2)

[false/q] $\neg(\wedge\Theta \wedge \neg q)$ iff (by propositional logic)

$\neg$[false/q] $\wedge\, \Theta$. $\dashv$

This rule may be checked in the preceding example. Indeed, we have that [false/q] Θ is equivalent to $\neg p \wedge \neg r$, whose negation is equivalent to our earlier outcome $p \vee r$.

This analysis suggests abductive variations that we did not consider before. For instance, we need not forbid all closures involving $\neg q$, but only those which involve $\neg q$ in final position (i.e., negated forms of the 'surprising fact' to be explained). There might be other, harmless occurrences of $\neg q$ on a branch emanating from the background theory Θ itself.

Example

Let $\Theta = \{q \rightarrow p, p \rightarrow q\}$. The tableaux $\mathcal{T}(\Theta)$ is as follows:

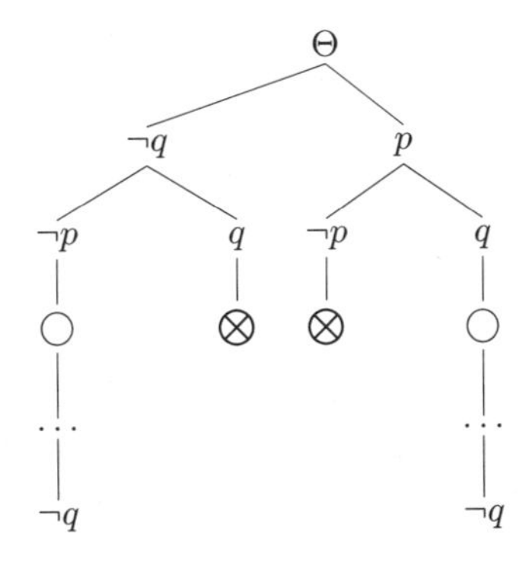

In this case, our earlier strategy would merely produce an outcome p - as can be checked by our false-computation rule. The new strategy, however, would compute an abduction p or q, which may be just as reasonable.

This analysis does not extend to complex conclusions. We leave its possible extensions open here.

Tableaux and Resolution

In spite of the logical equivalence between the methods of tableau and resolution [Fit90, Gal92], in actual implementations the generation of abductions turns out to be very different. The method of resolution used in logic programming does not handle negation explicitly in the language, and this fact restricts the kind of abductions to be produced. In addition, in logic programming only atomic formulas are produced as abductions since they are identified as those literals which make the computation fail. In semantic tableaux, on the other hand, it is quite natural to generate abductive formulas in conjunctive or disjunctive form as we have shown.

As for similarities between these two methods as applied to abduction, both frameworks have one of the explanatory conditions ($\Theta \not\Rightarrow \varphi$) already built in. In logic programming the abductive mechanism is put to work when a query fails, and in semantic tableaux abduction is triggered when the tableau for $\Theta \cup \neg\varphi$ is open.

Abduction in First Order Semantic Tableaux: The DB-Tableaux Model

The problem of abduction may be considered an impossible task when applied to first order logic [MP93]. The main reason being that the precondition for an abductive problem ($\mathcal{T}(\Gamma \cup \{\neg\phi\}$ has an open branch) is indeed undecidable. We cannot of course avoid that result but can tackle the problem for special cases.

In any case, if there is some presumption for modelling explanatory inference as conceived in the philosophy of science, quantified rules for dealing with universal laws must be incorporated. To this end, one problem to tackle is the existence of infinite branches for a given tableaux for a finitely satisfiable theory. An attractive way to do that would be to explore transformations of tableaux with infinite branches into others with finished ones; in such a way that the newly created finite branches would provide information that allows the construction of abductive explanations.

In this section the main aim is to study the problem of the existence of infinite branches when semantic tableaux are applied to the systematic search for a solution to an explanation problem. We propose the modification of the so called δ-rule (cf. below) into the δ'-rule (following previous work in [Dia93];

[Boo84]; [Nep 02] and [AN04]), in such a way that if a theory is finitely satisfiable, then the new semantic tableau presents the same advantages than the standard one to obtain certain explanatory facts.

In addition to the extension rules given for propositional tableaux (cf. subsection 2.1), here are the classical rules for first order tableaux:

- γ-rule:

$$\frac{\forall x \varphi}{\begin{array}{c}\varphi(b_1/x)\\ \varphi(b_2/x)\\ \ldots\\ \varphi(b_n/x)\end{array}}$$

  $b_1, ..., b_n$ are all constants that occur in sentences of the branch, $n \geq 1$ (if no one occurs, then $n = 1$), the new part of the branch is $\Phi + \varphi(b_1/x) + ... + \varphi(b_n/x)$ ($\varphi(b_i/x)$ is a γ-sentence, for all $i \leq n$).

- δ-rule

$$\frac{\exists x \varphi}{\varphi(b/x)}$$

  where b is a new constant that does not occur in any previous sentence of the branch, the new part of the branch is $\Phi + \varphi(b/x)$ and this is a δ-sentence.

We are interested in those quantified universal formulae of the following form: $\forall x \exists y \varphi$, which are satisfiable in finite domains. As it turns out, even formulae of this kind may have a tableau with infinite branches. Therefore, it is appealing to find a way by which it is possible to transform an infinite tableau into a finite one, in the sense of obtaining finite open branches.

Let us take as an example the following. The standard tableau for $\forall x \exists y Rxy$ is as follows:

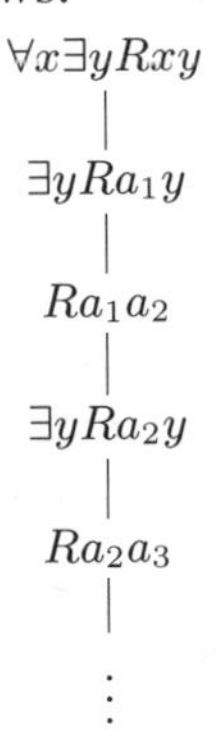

and so on. The only one branch is infinite, even though the formula is satisfiable in a unitary domain. Therefore, the classic tableau framework is of no use here. We will now introduce the DB-tableaux method[5], followed by the presentation of this same example, showing that a finite model is possible to be found.

DB-tableaux

Definition 1. *A DB-tableau is a tableau constructed by means of the usual extension rules for propositional tableaux, the γ-rule and the following one:*

δ'-rule:

$$\frac{\exists x \varphi}{\varphi(a_1/x) \mid \ldots \mid \varphi(a_{n+1}/x)}$$

Where $a_1, a_2, ..., a_n$ are all constants that occur in previous sentences; a_{n+1} is a new constant and $\varphi(a_i/x)$ is a δ'-sentence for every $i \leq n$. The branch Φ splits into the following branches: $\Phi + \varphi(a_1/x)$; $\Phi + \varphi(a_2/x)$; ...; $\Phi + \varphi(a_n/x)$; $\Phi + \varphi(a_{n+1}/x)$.

The DB-tableau for $\forall x \exists y Rxy$ is as follows:

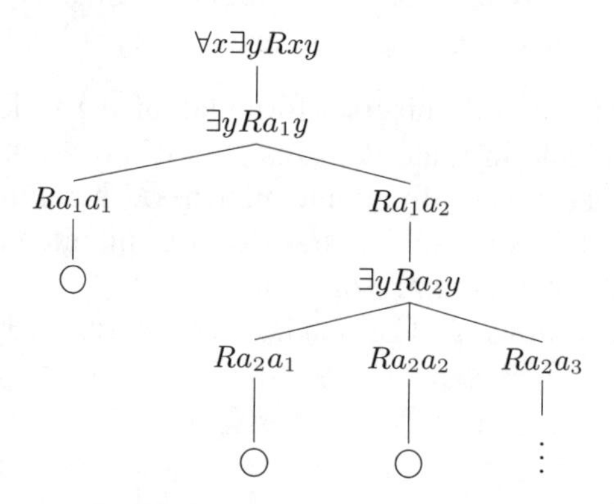

As illustrated, the DB-tableau for $\forall x \exists y Rxy$ has a completed open branch (in fact, three of them) from which a model $M = \langle D, \Im \rangle$ can be defined in such a way that M satisfies every formula of the branch, provided that $\Im(R) = \{\langle \Im(a_1), \Im(a_1) \rangle\}$.

[5] E. Díaz and G. Boolos proposed independently this modification of Beth's tableaux for the first time in [Dia93] and [Boo84] respectively.

In what follows we will introduce the notions of n-satisfiability and n-entailment for DB-tableaux in order to prepare the ground for our coming notion of an n-abductive problem.

Definition 2. *A branch of a DB-tableau has depth n, for $n \geq 1$, iff n is the number of (non repeated) constants occurring in the branch.*

Definition 3. *A finite set of sentences Γ is n-satisfiable, for $n \geqq 1$, iff there exists a domain D of cardinality n in which Γ is satisfiable.*

Definition 4. *Given a finite set of sentences Γ and a sentence ψ, Γ n-entails ψ, for $n \geqq 1$, in symbols: $\Gamma \models_n \psi$ iff $\Gamma \cup \{\psi\}$ is n-satisfiable and there is no domain of cardinality m, $m \leq n$, for which $\Gamma \cup \{\neg\psi\}$ is m-satisfiable.*

Theorem 1. *A finite set of sentences Γ is n-satisfiable, for $n \geq 1$ iff the DB-tableau of Γ has an open branch that is n in depth.*

This result is proved elsewhere (Cf. [Nep99] and [Nep 02]). An straightforward corollary of this theorem (together with definition 4) is the following theorem:

Theorem 2. *Given a set of sentences $\Gamma \subseteq L$ and a non quantified sentence $\psi \in L$, $\Gamma \models_n \psi$, for $n \geq 1$, iff the DB-tableau for $\Gamma \cup \{\psi\}$ has a completed open branch of depth n and $\Gamma \cup \{\neg\psi\}$ is not m-satisfiable, for every $m \leq n$.*

We have thus defined a notion of satisfaction and of entailment which are relative to domain cardinality. These notions are weaker than the standard ones, in fact, n-entailment may be labeled \weak entailment for n", given that now only some models (and not necessarily all) of Γ are models of ψ. As it should be noted, for every $n \geq 1$, if $\Gamma \models \psi$ then $\Gamma \models_n \psi$, though $\Gamma \models_n \psi$ does not imply $\Gamma \models \psi$. That is, Tarskian entailment implies n-entailment but not vice versa.

Moreover, this logical operation is non monotonic. For example, for $n = 2$, we have $\forall x \exists y Rxy, \neg Ra_2a_1 \models_2 Ra_2a_2$, but $\forall x \exists y Rxy, \neg Ra_2a_1, \forall x \neg Rxx \not\models_2 Ra_2a_2$. Here are the corresponding DB-tableaux for illustration:

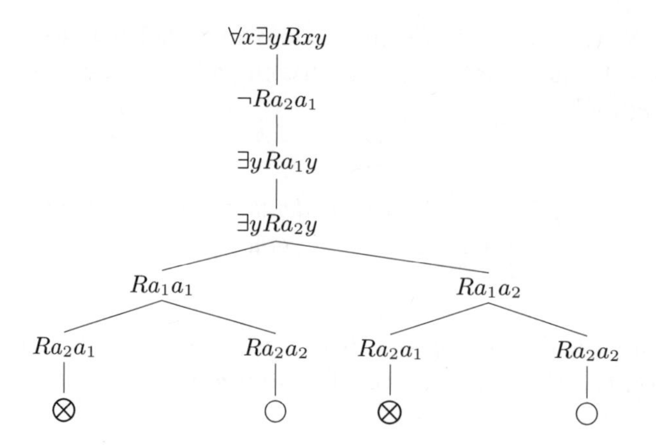

This tableau is closed by Ra_2a_2, which proves that the first entailment holds. The next tableau shows that the second entailment does not hold, for the branches of depth $n = 2$ close with the premisses alone.

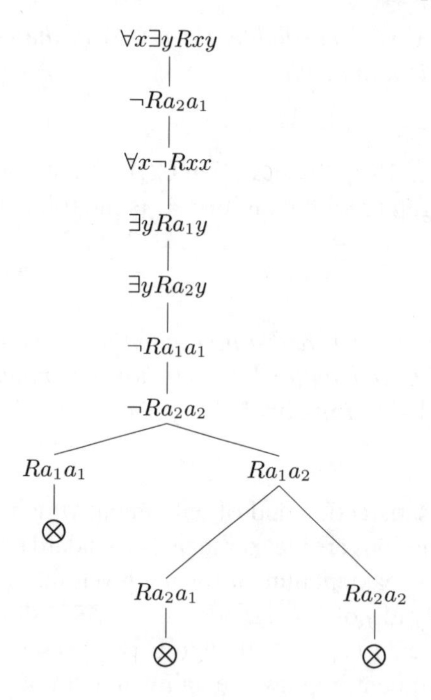

Abduction in First Order DB-Tableaux

In this section we will propose the use of DB-tableaux to work with some kind of abductive problems. Our approach so far has the limitation that produces only literals as abductive explanations, but the method could be extended to accommodate more complex forms as well (cf. [AN04]).

Definition 5. *Given a set of sentences $\Gamma \subseteq L$ and a non quantified sentence $\psi \in L$, for which the DB-tableau for $\Gamma \cup \{\neg\psi\}$ has an open branch of depth n, $\langle \Gamma, \psi \rangle_n$ is denoted as a n-abductive problem.*

Definition 6. *Given an n-abductive problem $\langle \Gamma, \psi \rangle_n$, $1 \leq n < \aleph_0$, α is a n-abductive solution iff*

1 *The DB-tableau of $\Gamma \cup \{\neg\psi\} \cup \{\alpha\}$ has closed every branch of depth m, for all $m \leq n$.*

2 *The DB-tableau of $\Gamma \cup \{\alpha\}$ has a completed open branch of depth n.*

These conditions correspond to those for the generating an abductive explanation in the classical model, as shown in section 3.2. However, now the two requirements are not relative to "all branches of the corresponding tableau", but rather relative to "all branches of the corresponding DB-tableau which are of depth n".

Example

Suppose that $\Gamma = \{\forall x \exists y Rxy\}$, and the fact to be explained is $\psi = \{Ra_1a_2\}$. We shall first construct the the DB-tableau for $\Gamma \cup \{\neg\psi\}$:

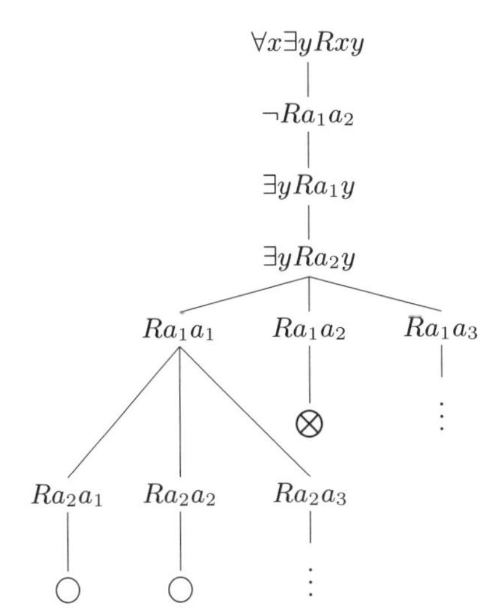

According to the first condition, a 2-abductive solution candidate is $\neg Ra_1a_1$, for this formula would close the above open branches of depth 2. Therefore, all branches of the DB-tableau for

$$\{\forall x \exists y Rxy, \neg Ra_1a_2, \neg Ra_1a_1\}$$

which are $m \leq 2$, are closed:

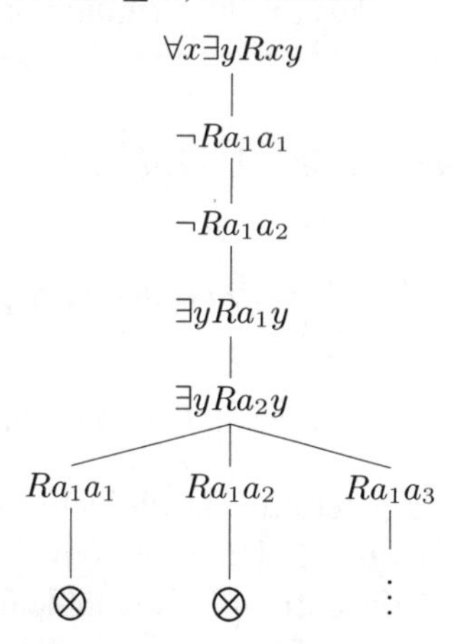

In fact, in this case there are no branches of smaller depth, since the formula in Γ has two variables, making it impossible to construct the tableau with branches of lesser depth. So we can conclude that:

$$\forall x \exists y Rxy, \neg Ra_1a_1 \models_2 Ra_1a_2$$

As for condition 2, the DB-tableau for $\{\forall x \exists y Rxy, \neg Ra_1a_1\}$ has a completed open branch (in fact two of them) of depth 2:

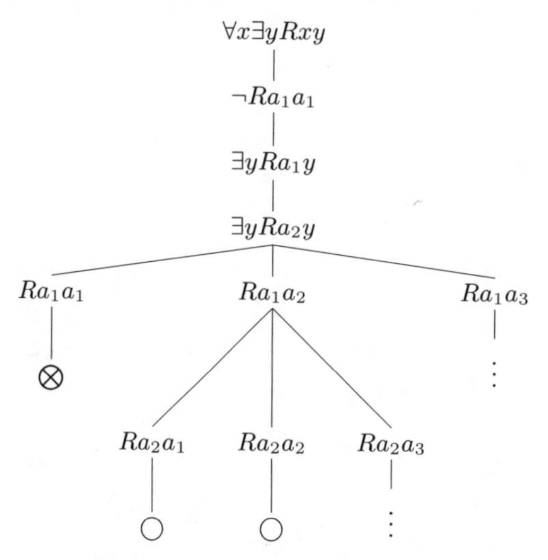

Therefore, $\neg Ra_1a_1$ is a 2-abductive solution for the abductive problem $\langle \forall x \exists y Rxy, Ra_1a_2 \rangle_2$.

In this section the main aim was to study the problem of the existence of infinite branches when semantic tableaux are applied to the systematic search for a solution to an abductive problem. An attractive way to deal with such a problem is to transform a tableau with infinite branches into another one with finished ones; in such a way that the newly created finite branches provide information that allows the construction of abductive explanations, our main concern. To this end, we proposed the modification of the standard δ-rule for the existential quantifier into the δ'-rule, and accordingly labelled our tableau method as DB-tableaux, in such a way that if a theory is finitely satisfiable, then the new semantic tableau presents the same advantages than the standard one to obtain certain explanatory facts.

We introduced the notions of n-satisfiability and n-entailment, both relative to a finite domain. As it turned out, one of our main results shows that a set of sentences is n-satisfiable if and only if its corresponding DB-tableau has an open branch of depth n. Therefore, we can define the notion of an n-abductive problem and its corresponding n-abductive solution for a DB-tableau of $\Gamma \cup \{\neg\phi\}$, as one with an open branch of depth n and α as its solution when the DB-tableau for $\Gamma \cup \{\neg\phi\} \cup \{\alpha\}$ has closed every branch of depth m (for all $m \leq n$) and for which the DB-tableau for $\Gamma \cup \{\alpha\}$ has a completed open branch of depth n.

Any theory with postulates that are formalized in this way is supposed to have an intended interpretation which corresponds to a model with a finite universe. And when it is n-satisfiable ($n \geq 1$), then it is m-satisfiable for every $m \geq n$ (pace the upward theorem of Löwenheim-Skolem). However, an advantage of DB-tableaux in this case is that the method obtains a minimal model, that is, a model with the least possible cardinality.

Our approach so far seems has the limitation of producing only literals as abductive explanations, but our method could be extended in a natural way in order to accommodate more complex forms as well (cf. [AN04]).

6. Discussion and Conclusions

Abduction in Tableaux

Exploring abduction as a form of computation gave us further insight into this phenomenon. Our concrete algorithms implement some earlier points from chapter 3, which did not quite fit the abstract structural framework. Abductions come in different degrees (atomic, conjunctive, disjunctive-conditional), and each abductive condition corresponds to new procedural complexity. In practice, though, it turned out easy to modify the algorithms accordingly. Indeed these findings reflect an intuitive feature of explanation. While it is sometimes difficult to describe what an explanation is in general, it may be easier to construct a set of explanations for a particular problem.

As for the computational framework, semantic tableaux are a natural vehicle for implementing abduction. They allow for a clear formulation of what counts as an abductive explanation, while being flexible and suggestive as to possible modifications and extensions. Derivability and consistency, the ingredients of consistent and explanatory abduction, are indeed a natural blend in tableaux, because we can manipulate open and closed branches with equal ease. Hence it is very easy to check if the consistency of a theory is preserved when adding a formula. (By the same token, this type of conditions on abduction appears rather natural in this light.)

Even so, our actual algorithms were more than a straight transcription of the logical formulations in chapter 3 (which might be very inefficient). Our computational strategy provided an algorithm which produces consistent formulas, selecting those which count as explanations, and this procedure turns out to be more efficient than the other way around. Nevertheless, abduction in tableaux has no unique form, as we showed by some alternatives. A final insight emerging from a procedural view of abduction is the importance of the background theory when computing explanations. Bad theories in combination with certain observation will produce bad explanations. Sophisticated computation cannot improve that.

Finally, we explored abduction in first order tableaux. Our approach has the advantage of simplifying the systematic search for a solution to certain abductive problems, by means of first order DB-tableaux, when constants occur in the corresponding formulae and the standard tableaux method is not effective to find a solution because of the generation of infinite branches. That is, for theories which are finitely satisfiable but that nevertheless generate infinite branches under Beth's tableau method.

Our tableau approach also has clear limitations. It is hard to treat notions of abduction in which $\Rightarrow$ is some non-standard consequence. In particular, with an underlying statistical inference, it is unclear how probabilistic entailments should be represented. Our computation of abductions relies on tableaux being open or closed, which represent only the two extremes of probable inference. We do have one speculation, though. The computation of what we called 'partial explanations' (which close some but not all open branches) might provide a notion of partial entailment in which explanations only make a fact less surprising, without explaining it in full. (Cf. [Tij97] for other approaches to 'approximative deduction' in abductive diagnostic settings.) As for other possible uses of the tableau framework, the case of abduction as revision was not addressed here. In a further chapter (8), we shall see that we do get further mileage in that direction, too.

Related Work

The framework of semantic tableaux has relatively recently been used beyond its traditional logical purposes, especially in computationally oriented approaches. One example is found in [Nie96], implementing 'circumscription'. In connection with abduction, semantic tableaux are used in [Ger95] to model natural language presuppositions (cf. chapter 2, abduction in linguistics). A better–known approach is found in [MP93], a source of inspiration for our work, which we proceed to briefly describe and compare. The following is a mere sketch, which cannot do full justice to all aspects of their proposal.

Mayer and Pirri's Tableau Abduction

Mayer and Pirri's article presents a model for computing 'minimal abduction' in propositional and first–order logic. For the propositional case, they propose two characterizations. The first corresponds to the generation of all consistent and minimal explanations (where minimality means 'logically weakest'; cf. chapter 3). The second generates a single minimal and consistent explanation by a non-deterministic algorithm. The first–order case constructs abductions by reversed skolemization, making use of unification and what they call 'dynamic herbrandization' of formulae. To give an idea of their procedures for generating explanations, we merely note their main steps: (1) construction of 'minimal closing sets', (2) construction of abductive solutions as literals which close all branches of those sets, (3) elimination of inconsistent solutions. The resulting systems are presented in two fashions, once as semantic tableaux, and once as sequent calculi for abductive reasoning. There is an elegant treatment of the parallelism between these two. Moreover, a predicate–logical extension is given, which is probably the first significant treatment of first–order abduction which goes beyond the usual resolution framework. (In subsequent work, the authors have been able to extend this approach to modal logic, and default reasoning.)

Comparison to our work

Our work has been inspired by [MP93]. But it has gone in different directions, both concerning strategy and output. (i) Mayer and Pirri compute explanations in line with version 1 of the general strategies that we sketched earlier. That is, they calculate all closures of the relevant tableau to later eliminate the inconsistent cases. We do the opposite, following version 2. That is, we first compute consistent formulas which close at least one branch of the original tableau, and then check which of these are explanations. Our reasons for this had to do with greater computational efficiency. (ii) As for the type of explanations produced, Mayer and Pirri's propositional algorithms basically produce minimal atomic explanations or nothing at all, while our approach provides explanations in conjunctive and disjunctive form as well. (iii) Mayer's and Pirri's approach stays closer to the classical tableau framework, while ours gives it several new twists. We propose several new distinctions and extensions, e.g. for the purpose of identifying when there are no consistent explanations at all. (iv) An interesting point of the Mayer and Pirri presentation is that it shows very well the contrast between computing propositional and first–order abduction. While the former is easy to compute, even considering minimality, the latter is inherently undecidable, but we have shown that even so there are interesting cases for treating abduction in first order tableaux. (iv) Eventually, we go even further (cf. chapter 8) and propose semantic tableaux as a vehicle for revision, which requires a new contraction algorithm.

PART III

APPLICATIONS

Chapter 5

SCIENTIFIC EXPLANATION

1. Introduction

In the philosophy of science, we confront our logical account of chapter 3 with the notion of scientific explanation, as proposed by Hempel in two of his models of scientific inference: deductive-nomological and inductive-statistical [Hem65]. We show that both can be viewed as forms of (abductive) explanatory arguments, the ultimate products of abductive reasoning. The former with deductive underlying inference, and the latter with statistical inference.

In this confrontation, we hope to learn something about the scope and limitations of our analysis so far. We find a number of analogies, as well as new challenges. The notion of statistical inference gives us an opportunity to expand the logical analysis at a point left open in chapter 3. We will also encounter further natural desiderata, however, which our current analysis cannot handle.

Our selection of topics in this field is by no means complete. We will cover some of them in the companion chapter to the philosophy of science (chapter 6), but many other connections exist relating our proposal to philosophy of science. Some of these concerns (such as cognitive processing of natural language) involve abductive traditions beyond the scope of our analysis. (We have already identified some of these in chapter 2.) Nevertheless, the connections that we do develop have a clear intention. We view all three fields as naturally related components in cognitive science, and we hope to show that abduction is one common theme making for cohesion.

2. Scientific Explanation as Abduction

At a general level, our discussion in chapters 1 and 2 already showed that scientific reasoning could be analyzed as an abductive process. This reasoning comes in different kinds, reflecting (amongst others) various patterns of discovery, with different 'triggers'. A discovery may be made to explain a novel phenomenon which is consistent with our current theory, but it may also point at an anomaly between the theory and the phenomenon observed. Moreover, the results of scientific reasoning vary in their *degree* of novelty and complexity. Some discoveries are simple empirical generalizations from observed phenomena, others are complex scientific theories introducing sweeping new notions. We shall concentrate on rather 'local' scientific explanations, which can be taken to be some sort of logical arguments: (abductive) explanatory ones, in our view. (We are aware of the existence of divergent opinions on this: cf. chapters 1 and 2). That scientific inference can be viewed as some form of abductive inference should not be surprising. Both Peirce and Bolzano were inspired in their logical systems by the way reasoning is done in science. Peirce's abductive formulations may be regarded as precursors of Hempel's notion of explanation, as will become clear shortly. Indeed, it has been convincingly claimed that *'Peirce should be regarded as the true founder of the theory of inductive-probabilistic explanation'* ([Nii81, p. 444] and see also [Nii00] for a detailed account).

At the center of our discussion on the logic of explanation lies the proposal by Hempel and Oppenheim [HO48, Hem65]. Their aim was to model explanations of empirical 'why-questions'. For this purpose they distinguished several kinds of explanation, based on the logical relationship between the explanans and explanandum (deductive or inductive), as well as on the form of the explanandum (singular events or general regularities). These two distinctions generate four models altogether: two deductive-nomological ones (D-N), and two statistical ones (Inductive-Statistical (I-S), and Deductive-Statistical (D-S)).

We analyze the two models for singular events, and present them as forms of abduction, obeying certain structural rules.

The Deductive-Nomological Model

The general schema of the D-N model is the following:

$$\frac{\begin{array}{l} L_1, \ldots, L_m \\ C_1, \ldots, C_n \end{array}}{E}$$

$L_1, \ldots, L_m$ are general laws which constitute a scientific theory T, and together with suitable antecedent conditions $C_1, \ldots, C_n$ constitute a *potential explanation* $\langle T, C \rangle$ for some observed phenomenon E. The relationship between explanandum and explananda is deductive, signaled by the horizontal line in the schema. Additional conditions are then imposed on the explananda:

$\langle T, C \rangle$ is a *potential explanation* of E iff

- T is essentially general and C is singular.
- E is derivable from T and C jointly, but not from C alone.

The first condition requires T to be a 'general' theory (having at least one universally quantified formula). A 'singular' sentence C has no quantifiers or variables, but just closed atoms and Boolean connectives. The second condition further constrains the derivability relation. Both T and C are required for the derivation of E.

Finally, the following empirical requirement is imposed:

$\langle T, C \rangle$ is an *explanation* of E iff

- $\langle T, C \rangle$ is a potential explanation of E
- C is true

The sentences constituting the explananda must be true. This is an empirical condition on the status of the explananda. $\langle T, C \rangle$ remains a *potential explanation* for E until C is verified.

From our logical perspective of chapter 3, the above conditions define a form of abduction. In potential explanation, we encounter the derivability requirement for the plain version ($T, C \vdash E$), plus one of the conditions for our 'explanatory' abductive style $C \not\vdash E$. The other condition that we had ($T \not\vdash E$) is not explicitly required above. It is implicit, however, since a significant singular sentence cannot be derived solely from quantified laws (which are usually taken to be conditional). An earlier major abductive requirement that seems absent is consistency ($T \not\vdash \neg C$). Our reading of Hempel is that this condition is certainly presupposed. Inconsistencies certainly never count as scientific explanations. Finally, the D-N account does not require minimality for explanations: it relocates such issues to choices between better or worse explanations, which fall outside the scope of the model. We have advocated the same policy for abduction in general (leaving minimal selection to our algorithms of chapter 4).

There are also differences in the opposite direction. Unlike our (abductive) explanatory notions of chapter 2, the D-N account crucially involves restrictions on the form of the explanantia. Also, the truth requirement is a major difference. Nevertheless, it fits well with our discussion of Peirce in chapter 2: an abducible has the status of a suggestion until it is verified.

It seems clear that the Hempelian deductive model of explanation is closely related to our proposal of (abductive) explanatory argument, in that it complies with most of the logical conditions discussed in chapter 3. If we fit D-N explanation into a deductive format, the first thing to notice is that laws and initial conditions play different roles in explanation. Therefore, we need a ternary format of notation: $T \mid C \Rightarrow_{HD} E$. A consequence of this fact is that this inference is non-symmetric ($T \mid C \Rightarrow_{HD} E \not\Leftrightarrow C \mid T \Rightarrow_{HD} E$). Thus, we keep T and C separate in our further discussion. Here is the resulting notion once more:

Hempelian Deductive-Nomological Inference $\Rightarrow_{HD}$

$T \mid C \Rightarrow_{HD} E$ iff

(i) $T, C \vdash E$

(ii) T, C is consistent

(iii) $T \not\vdash E, C \not\vdash E$

(iv) T consists of universally quantified sentences,

C has no quantifiers or variables.

Structural Analysis

We now analyze this notion once more in terms of structural rules. For a general motivation of this method, see chapter 3. We merely look at a number of crucial rules discussed earlier, which tell us what kind of explanation patterns are available, and more importantly, how different explanations may be combined.

Reflexivity

Reflexivity is one form of the classical Law of Identity: every statement implies itself. This might assume two forms in our ternary format:

$E \mid C \Rightarrow_{HD} E \qquad T \mid E \Rightarrow_{HD} E$

However, Hempelian inference rejects both, as they would amount to 'irrelevant explanations'. Given condition (iii) above, neither the phenomenon E should count as an explanation for itself, nor should the theory contain the observed phenomenon: because no explanation would then be needed in the first place. (In addition, left reflexivity violates condition (iv), since E is not a universal but an atomic formula.) Thus, Reflexivity in this way has no place in a structural account of explanation.

Monotonicity

Monotonic rules in scientific inference provide means for making additions to the theory or the initial conditions, while keeping scientific arguments valid. Although deductive inference by itself is monotonic, the additional conditions on $\Rightarrow_{HD}$ invalidate classical forms of monotonicity, as we have shown in detail in chapter 3. So, we have to be more careful when 'preserving explanations', adding a further condition. Unfortunately, some of the tricks from chapter 3 do not work in this case, because of our additional language requirements.

Moreover, outcomes can be tricky. Consider the following monotonicity rule, which looks harmless:

HD Monotonicity on the Theory (invalid):

$$\frac{T|A \Rightarrow_{HD} E \qquad T, B|D \Rightarrow_{HD} E}{T, B|A \Rightarrow_{HD} E}$$

This says that, if we have an explanation A for E from a theory, as well as another explanation D for the same fact E from a strengthened theory, then the original explanation will still work in the strengthened theory. This sounds convincing, but it will fail if the strengthened theory is inconsistent with A.

If we add an extra requirement, that is, $T, B|A \Rightarrow_{HD} F$, then we get a valid form of monotonicity:

HD Monotonicity on the Theory (valid):

$$\frac{T|A \Rightarrow_{HD} E \qquad T, B|D \Rightarrow_{HD} E \qquad T, B|A \Rightarrow_{HD} F}{T, B|A \Rightarrow_{HD} E}$$

This additional condition forces to require that the strengthened theory explain something else (F), in order to warrantee the consistency of the theory and explanation with the observation. However, we have not been able to find any convincing monotonicity rules when only two requirements are set!

What we learn here is a genuine limitation of the approach in chapter 3. With notions of D-N complexity, pure structural analysis may not be the best way to go. We might also just bring in the additional conditions *explicitly*, rather than encoding them in abductive sequent form. Thus, 'a theory may be strengthened in an explanation, provided that this happens consistently, and without implying the observed phenomenon without further conditions'. It is important to observe that the complexities of our analysis are no pathologies, but rather reflect the true state of affairs. They show that there is much more to the logic of Hempelian explanation than might be thought, and that this can be brought out in a precise manner. For instance, the failure of monotonicity rules means that one has to be very careful, *as a matter of logic*, in 'lifting' scientific explanations to broader settings.

Cut

The classical Cut rule allows us to chain derivations, and replace temporary assumptions by further premises implying them. Thus, it is essential to standard reasoning. Can explanations be chained? Again, there are many possible cut rules, some of which affect the theory, and some the conditions. We consider one natural version:

HD Cut (invalid):

$$\frac{T|A \Rightarrow_{HD} B \quad T|B \Rightarrow_{HD} E}{T|A \Rightarrow_{HD} E}$$

Our rain example of chapter 2 gives an illustration of this rule. Nimbostratus clouds (A) explain rain (B), and rain explains wetness (E), therefore nimbostratus clouds (A) explain wetness (E). But, is this principle generally valid? It almost looks that way, but again, there is a catch. In case A implies E by itself, the proposed conclusion does not follow. Again, we would have to add this constraint separately, in order to have a valid form of cut:

HD Cut (valid):

$$\frac{T|A \Rightarrow_{HD} B \quad T|B \Rightarrow_{HD} E \quad A \nvdash E}{T|A \Rightarrow_{HD} E}$$

This rule not only has an additional requirement but it actually involves two kinds of inference to make it work ($\Rightarrow_{HD}$ and $\vdash$). We will see later on that rules involving inductive inference do require this kind of combination. Moreover, there may be reservation to add this extra requirement as predicate logic is only semi-decidable, and making claims in the negative gives not warrantee to achieve them. Still, if we meet the requirements set in the premisses we get a valid form of monotonicity.

Logical Rules

A notion of inference with as many side conditions as $\Rightarrow_{HD}$ has, considerably restricts the forms of valid structural rules one can get. Indeed, this observation shows that there are clear limits to the utility of the purely structural rule analysis, which has become so popular in a broad field of contemporary logic. To get at least some interesting logical principles that govern explanation, we must bring in logical connectives. Here are some valid rules.

1 Disjunction of Theories

$$\frac{T_1 \mid C \Rightarrow E \qquad T_2 \mid C \Rightarrow E}{T_1 \vee T_2 \mid C \Rightarrow E}$$

2 Conjunction of two Explananda

$$\frac{T \mid C \Rightarrow E_1 \qquad T \mid C \Rightarrow E_2}{T \mid C \Rightarrow E_1 \wedge E_2}$$

3 Conjunction of Explanandum and Theory

$$\frac{T \mid C_1 \Rightarrow E \qquad T \mid C_2 \Rightarrow F}{T \mid C_1 \Rightarrow F \wedge E}$$

4 Disjunction of Explanans and Explanandum

$$\frac{T \mid C \Rightarrow E}{T \mid C \vee E \Rightarrow E}$$

5 Weakening Explanans by Theory

$$\frac{T, F \mid A \Rightarrow E}{T, F \mid F \rightarrow A \Rightarrow E}$$

The first two rules show that well-known classical inferences for disjunction and conjunction carry over to explanation. The third says that the explanandum can be strengthened with any consequence of the background theory. However, it is common to demand that self explanations should be excluded (i.e., T does not explain T). Then, it is also required that F is not equal to T. This has been one of the obstacles to earlier attempts to give necessary and sufficient conditions for valid explanatory arguments (see [Tuo73]). The fourth shows how explananda can be weakened 'up to the explanandum'. The fifth rule states that explananda may be weakened provided that the background theory can compensate for this. The last rule actually highlights a flaw in Hempel's models, which he himself recognized. It allows for a certain trivialization of 'minimal explanations', which might be blocked again by imposing further syntactic restrictions (see [Hem65, Sal90]). However, note that these rules do not necessarily obey the earlier syntactic restrictions of the D-N model, for they involve non-atomic formulae in the explanandum (rules 2 and 3) and formula, which may contain quantifiers as initial condition ($F \rightarrow A$ in rule 5). This situation suggests dropping these restrictions if we want to handle rules with connectives (but then, we would not speak of Hempel's model for singular events). Moreover, the complete DN model of explanation involves both singular statements and generalizations. So, for a further analysis, an account of structural rules with quantificational principles of predicate logic is needed.

More generally, it may be said that the D-N model has been under continued criticism through the decades after its emergence. No generally accepted formal model of deductive explanation exists. But at least we hope that our style of analysis has something fresh to offer in this ongoing debate: if only, to bring simple but essential formal problems into clearer focus.

The Inductive-Statistical Model

Hempel's I-S model for explaining particular events E has essentially the same form as the D-N model. The fundamental difference is the status of the laws. While in the D-N model, laws are universal generalizations, in the

I-S model they are statistical regularities. This difference is reflected in the outcomes. In the I-S model, the phenomenon E is only derived 'with high probability' $[r]$ relative to the explanatory facts:

$$\frac{\begin{array}{l} L_1, \ldots, L_m \\ C_1, \ldots, C_n \end{array}}{E} \text{ [r]}$$

In this schema, the double line expresses that the inference is statistical rather than deductive. This model retains all adequacy conditions of the D-N model. But it adds a further requirement on the statistical laws, known as *maximal specificity* (RMS). This requirement responds to a problem which Hempel recognized as *the ambiguity of I-S explanation.* As opposed to classical deduction, in statistical inference, it is possible to infer contradictory conclusions from consistent premises. One of our examples from chapter 2 demonstrates this.

The Ambiguity of I-S Explanation

Suppose that theory T makes the following statements. "Almost all cases of streptococcus infection clear up quickly after the administration of penicillin (L1). Almost no cases of penicillin-resistant streptococcus infection clears up quickly after the administration of penicillin (L2). Jane Jones had streptococcus infection (C1). Jane Jones received treatment with penicillin (C2). Jane Jones had a penicillin-resistant streptococcus infection (C3)." From this theory it is possible to construct two *contradictory arguments*, one explaining why Jane Jones recovered quickly (E), and the other one explaining its negation, why Jane Jones did not recover quickly ($\neg E$):

Argument 1

$$\frac{\begin{array}{l} L_1 \\ C_1, C_2 \end{array}}{E} \text{ [r]}$$

Argument 2

$$\frac{\begin{array}{l} L_2 \\ C_2, C_3 \end{array}}{\neg E} \text{ [r]}$$

The premises of both arguments are consistent with each other, they could all be true. However, their conclusions contradict each other, making these arguments rival ones. Hempel hoped to solve this problem by forcing all statistical laws in an argument to be *maximally specific*. That is, they should contain all relevant information with respect to the domain in question. In our example, then, premise C3 of the second argument invalidates the first argument, since the law L1 is not maximally specific with respect to all information about Jane in T. So, theory T can only explain $\neg E$ but not E.

The RMS makes the notion of I-S explanation relative to a knowledge situation, something described by Hempel as 'epistemic relativity'. This requirement helps, but it is neither a definite nor a complete solution. Therefore, it has remained controversial[1].

These problems may be understood in logical terms. Conjunction of Consequents was a valid principle for D-N explanation. It also seems a reasonable principle for explanation generally. But its implementation for I-S explanation turns out to be highly non-trivial. The RMS may be formulated semi-formally as follows:

Requirement of Maximal Specificity:

A universal statistical law $A \rightsquigarrow B$ is maximally specific iff for all A' such that $A' \subset A$, $A' \rightsquigarrow B$.

We should note however, that while there is consensus of what this requirement means on an intuitive level, there is no agreement as to its precise formalization (cf. [Sal90] for a brief discussion on this). With this caveat, we give a version of I-S explanation in our earlier format, presupposing some underlying notion $\Rightarrow_i$ of inductive inference.

Hempelian Inductive Statistical Inference $\Rightarrow_{HI}$

$T, C \Rightarrow_{HI} E$ iff

(i) $T, C \Rightarrow_i E$

(ii) T, C is consistent

(iii) $T \not\Rightarrow_i E, C \not\Rightarrow_i E$

(iv) T is composed of statistical quantified formulas (which may include forms like "Most A are B"). C has no quantifiers.

(v) RMS: All laws in T are maximally specific with respect to T,C.

The above formulation complies with our earlier abductive requirements, but the RMS further complicates matters. Moreover, there is another source of vagueness. Hempel's D-N model fixes predicate logic as its language for making form distinctions, and classical consequence as its underlying engine. But in the I-S model the precise logical nature of these ingredients is left unspecified.

---

[1]One of its problems is that it is not always possible to identify a maximal specific law given two rival arguments. Examples are cases where the two laws in conflict have no relation whatsoever, as in the following example, due to Stegmüller [Ste83]: Most philosophers are not millionaires. Most mine owners are millionaires. John is both a philosopher and a mine owner. Is he a millionaire?

Some Interpretations of $\Rightarrow_i$

Statistical inference $\Rightarrow_i$ may be understood in a qualitative or a quantitative fashion. The former might read $\Theta \Rightarrow_i \varphi$ as: *"φ is inferred from Θ with high probability"*, while the latter might read it as *"most of the Θ models are φ models"*. These are different ways of setting up a calculus of inductive reasoning[2].

In addition, we have similar options as to the language of our background theories. The general statistical statements $A \rightsquigarrow B$ that may occur in the background theory Θ may be interpreted as either *"The probability of B conditioned on A is close to 1"*, or as statements of the form: *"most of the A-objects are B-objects"*. We will not pursue these options here, except for noting that the last interpretation would allow us to use the theory of *generalized quantifiers*. Many structural properties have already been investigated for probabilistic generalized quantifiers (cf. [vLa96, vBe84b] for a number of possible approaches).

Finally, the statistical approach to inference might also simplify some features of the D-N model, based on ordinary deduction. In statistical inference, the D-N notion of non-derivability need not be a negative statement, but rather a *positive* one of inference with (admittedly) *low probability*. This interpretation has some interesting consequences that will be discussed at the end of this chapter. It might also decrease complexity, since the notion of explanation becomes more uniformly 'derivational' in its formulation.

Structural Analysis, Revisited

Again, we briefly consider some structural properties of I-S explanation. Our discussion will be informal, since we do not fix any formal explanation of the key notions discussed before.

As for Reflexivity of $\Rightarrow_i$, this principle fails for much the same reasons as for D-N explanation. Next, consider Monotonicity. This time, new problems arise for strengthening theories in explanations, due to the RMS. A law L might be maximally specific with respect to $T|A$, but not necessarily so with respect to $T, B|A$ or to $T|A, B$. Worse still, adding premises to a statistical theory may reverse previous conclusions! In the above penicillin example, the theory without C3 explains perfectly why Jane Jones recovered quickly. But adding C3 reverses the inference, explaining instead why she did not recover quickly. If we then add that she actually took some homeopathic medicine with high chances of recovery (cf. chapter 2), the inference reverses again, and we will be able to explain once more why she recovered quickly.

[2] Recall from chapter 3 that there may be two different characterizations of the qualitative kind, one in which it is improbable that $\Theta \wedge \neg\varphi$ while another one in which given Θ, it is improbable that $\neg\varphi$.

These examples show once again that inductive statistical explanations are epistemically relative. There is no guarantee of preserving consistency when changing premises, therefore making monotonicity rules hopeless. And stating that the additions must be 'maximally specific' seems to beg the question. Nevertheless, we can salvage some monotonicity, provided that we are willing to *combine* deductive and inductive explanations. (There is no logical reason for sticking to pure formulations.) Here is a principle that we find plausible, modulo further specification of the precise inductive inference used (recall that $\Rightarrow_{HD}$ stands for deductive–nomological inference and $\Rightarrow_{HI}$ for inductive–statistical):

Monotonicity on the theory:

$$\frac{T|A \Rightarrow_{HI} C \qquad T, B|D \Rightarrow_{HD} C}{T, B|A \Rightarrow_{HI} C}$$

This rule states that statistical arguments are monotonic in their background theory, at least when what is added explains the relevant conclusion deductively with some other initial condition. This would indeed be a valid formulation, provided that we can take care of consistency for the enlarged theory $T, B \mid A$. In particular, all maximal specific laws for $T \mid A$ remain specific for $T, B \mid A$. For, by inferring C in a deductive and consistent way, there is no place to add something that would reverse the inference, or alter the maximal specificity of the rules.

Here is a simple illustration of this rule. If on the one hand, Jane has high chances of recovering quickly from her infection by taking penicillin, and on the other she would recover by taking some medicine providing a sure cure (B), she still has high chances of recovering quickly when the assertion about this cure is added to the theory.

Not surprisingly, there is no obvious Cut rule in this setting either. Here it is not the RMS causing the problem, as the theory does not change, but rather the well-known fact that statistical implications are not transitive. Again, we have a proposal for a rule combining statistical and deductive explanation, which might form a substitute: The following is our proposed formulation:

Deductive cut on the explanans:

$$\frac{T|A \Rightarrow_{HI} B \qquad T|B \Rightarrow_{HD} C}{T|A \Rightarrow_{HI} C}$$

Again, here is a simple illustration of this rule. If the administration of penicillin does explain the recovery of Jane with high probability, and this in turn explains deductively her good mood, penicillin explains with high probability Jane's good mood. (This reflects the well-known observation that "most A are B, all B are C" implies "most A are C". Note that the converse chaining is invalid.)

Patrick Suppes (p.c.) has asked whether one can formulate a more general representation theorem, of the kind we gave for consistent abduction (cf.

chapter 3), which would leave room for statistical interpretations. This might account for the fluid boundary between deduction and induction in common sense reasoning. Exploring this question however, has no unique and easy route. As we have seen, it is very hard to formulate a sound monotonic rule for I-S explanation. But this failure is partly caused by the fact that this type of inference requires too many conditions (the same problem arose in the H-D model). So, we could explore a calculus for a simpler notion of probabilistic consequence, or rather work with a combination of deductive and statistical inferences. Still, we would have to formulate precisely the notion of probabilistic consequence, and we have suggested there are several (qualitative and quantitative) options for doing it. Thus, characterizing an abductive logical system that could accommodate statistical reasoning is a question that deserves careful analysis, beyond the confines of this book.

3. Discussion and Conclusions

Further Relevant Connections

In this section we briefly present the notion of inductive support, show its relation to the notion of explanation and attempt to hint at some of its structural properties. Moreover, we review Salmon's notion of statistical relevance, as a sample of more sophisticated post-Hempelian views of explanation. In addition, we briefly discuss a possible computational implementation of Hempel's models more along the lines of our chapter 4, and finally we present Thagard's notion of explanation within his account of 'computational philosophy of science'.

Inductive Support

Related to the notion of explanation is the notion of *inductive support*. Taking into account some claimed logical principles around this notion we can explore some of its features under the light of our structural analysis given so far.

Intuitively, 'e inductively supports h', referred to as eIh, if h is 'inducible' from e, something which presupposes some inductive consequence logical relation between e and h. Some relevant principles are the following (see [NT73] for motivation and other principles):

- (SC) Special Consequence: If eIh and $h \vdash b$, then eIb.
- (CC) Converse Consequence: If eIh and $b \vdash h$, then eIb.
- (E) Entailment: If $e \vdash h$, then eIh.

Along the lines of chapter 3, we can capture such inductive logical relationship (eIh) in its forward fashion ($h \Rightarrow_{is} e$), for some suitable notion of $\Rightarrow_{is}$. Then a principle like (CC) may be expressed as follows:

- (CC) If $h \Rightarrow_{is} e$ and $b \vdash h$, then $b \Rightarrow_{is} e$.

It is easy to see that using this principle we can conclude $b, h \Rightarrow_{is} e$ from $h \Rightarrow_{is} e$ and $b \vdash h$. We have just to appeal to the monotonicty of $\vdash$ (from $b \vdash h$ we can infer $b, h \vdash h$). However, it this seems incorrect to infer that e inductively supports b, h from the fact that e alone inductively supports h, even if we have that h is deductively inferred from b. Thus, a revision of this principle is proposed as follows [Bro68]:

- (CC*) If $h \Rightarrow_{is} e$ and 'b explains h', then $b \Rightarrow_{is} e$.

This version blocks the previous undesired monotonicity, for the relationship of explanation is not preserved when the explananda is strengthened with additional formulae (see [NT73, p. 225] for a discussion of the adoption of this principle). Thus, it seems clear that notions like explanation and confirmation do not obey the classical monotonicty rule and therefore appropriate formulations have to be put forward. We leave this as an open question for the case of confirmation. The purpose of this brief presentation was just to show the possibility of studying this notion at a structural level, as we have done in chapter 3 for (abductive) explanatory arguments.

Salmon's Statistical Relevance

Despite the many conditions imposed, the D-N model still allows explanations irrelevant to the explanandum. The problem of characterizing when an explanans is *relevant* to an explanandum is a deep one, beyond formal logic. It is a key issue in the general philosophy of science.

One noteworthy approach regards relevance as *causality*. W. Salmon first analyzed explanatory relevance in terms of statistical relevance [Sal71]. For him, the issue in inductive statistical explanation is not how probable the explanans $(T|A)$ renders the explanandum C, but rather whether the facts cited in the explanans make a difference to the probability of the explanandum. Thus, it is not high probability but statistical relevance that makes a set of premises statistically entail a conclusion.

Now recall that we said that statistical non-derivability ($\Theta \not\Rightarrow_i \varphi$) might be refined to state that "φ follows with low probability from Θ". With this reinterpretation, one can indeed measure if an added premise changes the probability of a conclusion, and thereby count as a relevant explanation. This is not all there is to causality. Salmon himself found problems with his initial proposal, and later developed a causal theory of explanation [Sal84], which was refined in [Sal94]. Here, explanandum and explananda are related through a *causal nexus* of causal processes and causal interactions. Even this last version is still controversial (cf. [Hit95]).

A Computational Account of Hempel's models?

Hempel treats scientific explanation as a given product, without dealing with the processes that produce such explanations. But, just as we did in chapter 4, it seems natural to supplement this inferential analysis with a computational search procedure for producing scientific explanations. Can we modify our earlier algorithms to do such a job? For the D-N model, indeed, easy modifications to our algorithm would suffice. (But for the full treatment of universal laws, we would need a predicate-logical version of the tableau algorithm in line with our proposal of DB-tableaux in chapter 4.) For the inductive statistical case, however, semantic tableaux seem inappropriate. We would need to find a way of representing statistical information in tableaux, and then characterize inductive inference inside this framework. Some more promising formalisms for computing inductive explanations are labeled deductive systems with 'weights' by Dov Gabbay [Gab96], and other systems of dependence-driven qualitative probabilistic reasoning by W. Meyer Viol (cf. [Mey95]) and van Lambalgen & Alechina ([AL96]).

Thagard's Computational Explanation

An alternative route toward a procedural account of scientific explanation is taken by Thagard in [Tha88]. Explanation cannot be captured by a Hempelian syntactic structure. Instead, it is analyzed as a *process of providing understanding*, achieved through a mechanism of 'locating' and 'matching'. The model of explanation is the program **PI** ('processes of induction'), which computes explanations for given phenomena by procedures such as abduction, analogy, concept formation, and generalization and then accounts for a 'best explanation' comparing those which the program is able to construct.

This approach concerns abduction in cognitive science (cf. chapter 2), in which explanation is regarded as a problem–solving activity modeled by computer programs. (The style of programming here is quite different from ours in chapter 4, involving complex modules and record structures.) This view gives yet another point of contact between contemporary philosophy of science and artificial intelligence.

Explanation in Belief Revision

We conclude with a short discussion that relates abduction with scientific explanation. Theories of explanation in the philosophy of science mainly concern scientific inquiry. Nevertheless, some ideas by Gärdenfors on explanation (chapter 8 of his book [Gar88]), turn out illuminating in creating a connection. Moreover, how explanations are computed for incoming beliefs makes a difference in the type of operation required to incorporate the belief.

Gärdenfors' basic idea is that an explanation is, that which makes the explanandum less surprising by raising its probability. The relationship between explananda and explanandum is relative to an epistemic state, based on a probabilistic model involving a set of possible worlds, a set of probability measures, and a belief function. Explanations are propositions that effect a special epistemic change, increasing the belief value of the explanandum. Explananda must also convey information that is 'relevant' to the beliefs in the initial state. This proposal is very similar to Salmon's views on statistical relevance, which we discussed in connection with our statistical reinterpretation of non-derivability as derivability with low probability. The main difference between the two is this. While Gärdenfors requires that the change is in raising the probability, Salmon admits just any change in probability. Gärdenfors notion of explanation is closer to our 'partial explanations' of chapter 4 (which closed some but not all of the available open tableau branches). A natural research topic will be to see if we can implement the idea of raising probability in a qualitative manner in tableaux. Gärdenfors' proposal involves degrees of explanation, suggesting a measure for explanatory power with respect to an explanandum.

Combining explanation and belief revision (cf. chapter 8) into one logical endeavor also has some broader attractions. This co-existence (and possible interaction) seems to reflect actual practice better than Hempel's models, which presuppose a view of scientific progress as mere accumulation. This again links up with philosophical traditions that have traditionally been viewed as a non–logical, or even anti-logical. These include historic and pragmatic accounts [Kuh70, vFr80] that focus on analyzing explanations of anomalous instances as cases of *revolutionary* scientific change. ("Scientific revolutions are taken to be those noncumulative developmental episodes in which an older paradigm is replaced in whole or in part by an incompatible new one" [Kuh70, p.148].) In our view, the Hempelian and Kuhnian schools of thought, far from being enemies, emphasize rather two sides of the same coin.

Finally, our analysis of explanation as compassing both scientific inference shows that the philosophy of science and artificial intelligence share central aims and goals. Moreover, these can be pursued with tools from logic and computer science, which help to clarify the phenomena, and show their complexities.

Conclusions

Our analysis has tested the logical tools developed in chapter 3 on Hempel's models of scientific explanation. The deductive-nomological model is indeed a form of (abductive) explanatory inference. Nevertheless, structural rules provide limited information, especially once we move to inductive-statistical explanation. In discussing this situation, we found that the proof theory of combinations of D-N and I-S explanation may actually be better-behaved than either by itself. Even in the absence of spectacular positive insights, we can still observe that the abductive inferential view of explanation does bring it in line with modern non-monotonic logics.

Our negative structural findings also raise interesting issues by themselves. From a logical point of view, having an inferential notion like $\Rightarrow_{HD}$ without reflexivity and monotonicity, challenges a claim made in [Gab85]. Gabbay argues that reflexivity, 'restricted mononoonicity' and cut are the three minimal conditions which any consequence relation should satisfy to be a *bona fide* non-monotonic logic. (Later in [Gab94a] however, Gabbay admits this is not such a strong approach after all, as *'Other systems, such as relevance logic, do not satisfy even reflexivity'*.) We do not consider this failure an argument against the inferential view of explanation. As we have suggested in chapter 3, there are no universal logical structural rules that fit every consequence relation. What counts rather is that such relations 'fit a logical format'.

Non-monotonic logics have been mainly developed in AI. As a further point of interest, we mention that the above ambiguity of statistical explanation also occurs in default reasoning. The famous example of 'Quakers and Pacifists' is just another version of the one by Stegmüller cited earlier in this chapter. More specifically, one of its proposed solutions by Reiter [Rei80] is in fact a variation of the RMS[3]. These cases were central in Stegmüller criticisms of the positivist view of scientific explanation. He concluded that there is no satisfactory analysis using logic. But these very same examples are a source of inspiration to AI researchers developing non-standard logics.

There is a lot more to these connections than what we can cover in this book. With a grain of salt, contemporary logic-based AI research may be viewed as logical positivism 'pursued by other means'. One reason why this works better nowadays than in the past, is that Hempel and his contemporaries thought that classical logic was *the logic* to model and solve their problems. By present lights, it may have been their logical apparatus more than their research program what hampered them. Even so, they also grappled with significant problems, which are as daunting to modern logical systems as to classical ones, including a variety of pragmatic factors. In all, there seem to be ample reasons for philosophers of science, AI researchers, and logicians for making common cause.

Analyzing scientific reasoning in this broader perspective also sheds light on the limitations of this book. We have not really accounted for the distinction between laws and individual facts, we have no good account of 'relevance', and we have not really fathomed the depths of probability. Moreover, much of scientific explanation involves *conceptual change*, where the theory is modified with new notions in the process of accounting for new phenomena. So far,

[3]In [Tan92] the author shows this fact in detail, relating current default theories to Hempel's I-S model.

neither our logical framework of chapter 3, nor our algorithmic one of chapter 4 has anything to offer in this more ambitious realm.

Chapter 6

EMPIRICAL PROGRESS

1. Introduction

Traditional positivist philosophy of science inherits from logical research not only its language, but also its focus on the *truth question*, that is to say, the purpose of using its methods as means for testing hypotheses or formulae. As we saw in the previous chapter, Hempelian models of explanation and confirmation seek to establish the conditions under which a theory (composed by scientific laws) together with initial conditions, explains a certain phenomenon or whether certain evidence confirms a theory. As for logical research, it has been characterized by two approaches, namely the syntactic and the semantic. The former account characterizes the notion of derivability and aims to answer the following question: given theory H (a set of formulae) and formula E, is E derivable from H? The latter characterizes the notion of logical consequence and responds to the following question: is E a logical consequence of H? (Equivalent to: are the models of H models of E?) Through the truth question we can only get a "yes–no" answer with regard to the truth or falsity of a given theory. Aiming solely at this question implies a static view of scientific practice, one in which there is no place for theory evaluation or change. Notions like derivation, logical consequence, confirmation, and refutation are designed for the corroboration –logical or empirical– of theories.

However, a major concern in philosophy of science is also that of theory evaluation. Since the 1930s, both Carnap and Popper stressed the difference between truth and confirmation, where the last notion concerns theory evaluation as well. Issues like the internal coherence of a theory as well as its confirming evidence, refuting anomalies or even its lacunae, are all key considerations to evaluate a scientific theory with respect to itself and existing others. While there is no agreement about the precise characterization of theory evaluation, and in general about what progress in science amounts to, it is clear that these notions go much further than the truth question. The responses that are searched for are answers to the *success question*, which includes the set of successes, failures and lacunae of a certain theory. This is interesting and useful beyond testing and evaluation purposes, since failures and lacunae indicate the problems of a theory, issues that if solved, would result in an improvement of a theory to give a better one. Thus, given the success question, the *improvement question* follows from it.

For the purposes of this chapter, we take as our basis a concrete proposal which aims at modeling the three questions about empirical progress set so far, namely the questions of truth, success and improvement, the approach originally presented by Theo Kuipers in [Kui00] and in [Kui99]. In particular, we undertake the challenge of the latter publication, namely to operationalize the task of theory revision aiming at empirical progress, that is, the task of *instrumentalist abduction*. Our proposal aims at showing that evaluation and improvement of a theory can be modeled by (an extension of) the framework of semantic tableaux, in the style of chapter 4. In particular, we are more interested in providing a formal characterization of *lacunae*, namely those phenomena which a scientific theory cannot explain, but that are consistent with it. This kind of evidence completes the picture of successes and failures of a theory [Kui99]. The terms 'neutral instance' and 'neutral result' are also used [Kui00], for the cases in which a(n) (individual) fact is compatible with the theory, but not derivable from it. This notion appears in other places in the literature. For example, Laudan [Lau77] speaks of them as a 'non-refuting anomalies', to distinguish them from the real 'refuting anomalies'. We also find these kinds of statements in epistemology. Gärdenfors [Gar88] characterizes three epistemic states of an agent with respect to a statement as follows: (i) acceptance, (ii) rejection, and (iii) undetermination, the latter one corresponding to the epistemic state of an agent in which she neither accepts nor rejects a belief, but simply has no opinion about it (cf. chapter 8 for a treatment of abduction as epistemic change based on this proposal). As for myself (cf. chapters 2 and 8), I have characterized "novelty abduction" as the epistemic operation that is triggered when faced with a 'surprising fact' (which is equivalent to the state of undetermination when the theory is closed under logical consequence). It amounts to the extension of the theory into another one that is able to explain the surprise.

These kind of phenomena put in evidence, not the falsity of the theory (as it is the case with anomalies), but its *incompleteness* ([Lau77]), its incapability of solving problems that it should be able to do.

In contemporary philosophy of science there is no place for *complete theories*, if we take a 'complete scientific theory' to mean that it has the capacity to give explanations –negative or positive– to all phenomena within its area of competence[1]. Besides, in view of the recognition of the need for initial conditions to account for phenomena, the incompleteness of theories is just something implicit. Therefore, we should accept the property of being "incomplete" as a virtue of scientific theories, rather than a drawback as in the case of mathematics. A scientific theory will always encounter new phenomena that need to be accounted for, for which the theory in question is insufficient. However, this is not to say that scientists or philosophers of science give up when they are faced with a particular "undecidable phenomenon".

Our suggestion is that the presence of lacunae marks a condition in the direction of progress of a theory, suggesting an extension (or even a revision) of the theory in order to "decide" about a certain phenomenon which has not yet been explained, thus generating a "better" theory, one which solves a problem that the original one does not. Therefore, lacunae not only play a role in the evaluation of a theory, but also in its design and generation.

As for the formal framework for modeling the success and the improvement questions with regard to lacunae, classical methods in logic are no use. While lacunae correspond to "undecidable statements"[2], classical logic does not provide a method for modifying a logical theory in order to "resolve" undecidable statements. However, extensions to classical methods may provide suitable frameworks for representing theory change. We propose an extension to the method of semantic tableaux (cf. chapter 4), originally designed as a refutation procedure, but used here beyond its purpose in theorem proving and model checking, as a way to extend a theory into a new one which entails the lacunae of the original one. This is a well-motivated standard logical framework, but over these structures, different search strategies can model empirical progress in science.

This chapter is naturally divided into four parts. After this introduction, in the second part (section 2) we describe Kuipers' empirical progress framework, namely the characterization of evidence types, the account of theory evaluation and comparison, as well as the challenge of the instrumentalist abduction task. In the third part (section 3) we present our approach to empirical progress in the (extended) semantic tableaux framework. In the final part of this chapter (section 4), we sum up our previous discussion and offer our general conclusions.

[1]It is clear that a theory in physics gives no account to problems in molecular biology. Explanation here is only within the field of each theory. Moreover, in our interpretation we are neglecting all well-known problems that the identification between the notions of explanation and derivability brings about, but this understanding of a 'complete scientific theory' highlights the role of lacunae, as we will see later on.

[2]In 1931 Gödel showed that if we want a theory free of contradictions, then there will always be assertions we cannot decide, as we are tied to incompleteness. We may consider undecidable statements as extreme cases of lacunae, since Gödel did not limit himself to show that the theory of arithmetic is incomplete, but in fact that it is "incompletable", by devising a method to generate undecidable statements.

2. Kuipers' Empirical Progress

Two basic notions are presented by this approach, which characterize the relationship between scientific theories and evidence, namely confirmation and falsification. To this end, Kuipers proposes the *Conditional Deductive Confirmation Matrix*. For the purposes of this chapter, we are using a simplified and slightly modified version of Kuipers' matrix. On the one hand, we give the characterization by evidence type, that is, successes and failure, rather than by the notions of confirmation and falsification, as he does. Moreover, we omit two other notions he characterizes, namely (dis)confirmation and verification. On the other hand, we do not take account of the distinction between general successes and individual problems (failures), something that is relevant for his further characterization of theory improvement. Our version is useful since it gives lacunae a special place in the characterization, on a par with successes and failures, and it also permits a quite simple representation in our extended tableaux framework.

Success and Failure

SUCCESS

Evidence E is a success of a theory H relative to an initial condition C whenever:

$H, C \models E$

In this case, E *confirms* H relative to C.

FAILURE

Evidence E is a failure of a theory H relative to an initial condition C whenever:

$H, C \models \neg E$

In this case E *falsifies* H relative to C.

This characterization has the classical notion of logical consequence ($\models$) as the underlying logical relationship between a theory and its evidence[3]. Therefore, successes of a theory (together with initial conditions) are its logical consequences and failures are those formulae for which their negations are logical consequences. Additional logical assumptions have to be made. H and C must be logically independent, C and E must be true formulae and neither C nor H give an account of E for the case of confirmation (or of $\neg E$ for falsification) by themselves. That is:

1 Logical Independence

$H \not\models C, H \not\models \neg C$

$\neg H \not\models C, \neg H \not\models \neg C$

[3]But this need not to be so. Several other notions of semantic consequence (preferential, dynamic) or of derivability (default, etc.) may capture evidence type within other logical systems (cf. chapter 3 for a logical characterization of abduction within several notions of consequence).

2 C and E are true formulae.

3 Assumptions for Conditional Confirmation:
$H \not\models E, C \not\models E$

4 Assumptions for Conditional Falsification:
$H \not\models \neg E, C \not\models \neg E$

Notice that the requirement of logical independence assures the consistency between H and C, a requirement that has sometimes been overlooked (especially by Hempel, cf. chapter 5), but it is clearly necessary when the logical relationship is that of classical logical entailment. The assumptions of conditional confirmation and conditional falsification have the additional effect of preventing evidence from being regarded as an initial condition $(H, E \models E)$[4].

Lacuna

Previous cases of success and failure do not exhaust all possibilities there are in the relationship amongst H, C and E $(\neg E)$. It may very well be the case that given theory H, evidence E and all previous assumptions, there is no initial condition C available to account for E $(\neg E)$ as a case of success (or failure). In this case we are faced with a *lacuna* of the theory. More precisely:

LACUNA
Evidence E is a lacuna of a theory H when the following conditions hold for all available initial conditions C:
$H, C \not\models E$
$H, C \not\models \neg E$
In this case, E neither confirms nor falsifies H.

It is clear that the only case in which a theory has no lacuna is given when it is *complete*. But as we have seen, there is really no complete scientific theory. To find lacunae in a theory suggests a condition (at least from a logical point of view) in the direction of theory improvement. That is, in order to improve a theory (H_1) to give a better one (H_2), we may extend H_1 in such a way that its lacunae become successes of H_2. The existence of lacunae in a theory confront us with its holes, and their resolution into successes indicates progress in the theory. By so doing, we are *completing* the original theory (at least with respect

[4]These additional assumptions are of course implicit for the case in which E $(\neg E)$ is a singular fact (a literal, that is an atom or a negation thereof), since a set of universal statements H cannot by itself entail singular formulae, but we do not want to restrict our analysis to this case, as evidence in conditional form may also be of interest.

to certain evidence) and thus constructing a better one, which may include additional laws or initial conditions.

The above characterization in terms of successes, failures and lacunae makes it clear that the status of certain evidence is always with respect to some theory and specific initial condition. Therefore, in principle a theory gives no account of its evidence as successes or failures, but only with respect to one (or more) initial condition(s). Lacunae, on the other hand, show that the theory is not sufficient, that there may not be enough laws to give an account of evidence as a case of success or failure. But then a question arises: does the characterization of lacunae need a reference to the non-existence of appropriate initial conditions? In other words, is it possible to characterize lacuna-type evidence in terms of some condition between H and E alone? In the next subsection we prove that it is possible to do so in our framework.

Theory Evaluation and Comparison

Kuipers argues for a 'Context of Evaluation', as a more appropriate way to refer to the so-called 'Context of Justification' (cf. chapter 1):

> *'Unfortunately, the term 'Context of Justification', whether or not specified in a falsificationist way, suggests, like the terms 'confirmation' and 'corroboration', that the truth or falsity of a theory is the sole interest. Our analysis of the HD-method makes it clear that it would be much more adequate to speak of the* 'Context of Evaluation'. *The term 'evaluation' would refer, in the first place, to the separate and comparative HD-evaluation of theories in terms of successes and problems'.* [Kui00, p.132]

Kuipers proposes to extend Hempel's HD methodology to account not only for testing a theory, which only gives an answer to the 'truth-question', but also for evaluation purposes, in which case it allows one to answer the 'success-question', in order to evaluate a theory itself, thus setting the ground for comparing it with others. This approach shows that a further treatment of a theory may be possible after its falsification, it also allows a record to be made of the lacunae of a theory and by so doing it leaves open the possibility of improving it. As Kuipers rightly states, HD-testing leads to successes or problems (failures) of a certain theory 'and not to neutral results' [Kui00, 101].

There are, however, several models for HD-evaluation, including the *symmetric* and the *asymmetric* ones (each one involving a macro or a micro argument). For the purposes of this chapter we present a simplified version (in which we do not distinguish between individual and general facts) of the symmetric definition, represented by the following *comparative evaluation matrix* ([Kui00, 117]):

$H_2 \setminus H_1$	$Failures$	$Lacunae$	$Successes$
$Failures$	0	$-$	$-$
$Lacunae$	$+$	0	$-$
$Successes$	$+$	$+$	0

Besides neutral results (0), there are three types of results which are favorable (+) for H_2 relative to H_1, and three types of results which are unfavorable (-) for H_2 relative to H_1.

This characterization results in the following criterion for theory comparison, which allows one to characterize the conditions under which a theory is more successful than another: 'Theory H_2 is more successful than theory H_1 if there are, besides neutral results, some favorable results for H_2 and no unfavorable results for H_2'. Of course, it is not a question of counting the successes (or failures) of a theory and comparing it with another one, rather it is a matter of inclusion, which naturally leads to the following characterization: (cf. [Kui00, 112])

Theory Comparison

A theory H_2 is (at time t) as successful as (more successful than) theory H_1 if and only if (at time t):

1 The set of failures in H_2 is a subset of the set of failures in H_1.

2 The set of successes in H_1 is a subset of the set of successes in H_2

3 At least in one of the above cases the relevant subset is proper.

This model puts forward a methodology in science, which is formal in its representation, and it does not limit itself to the truth question. Nevertheless, it is still a static representation of scientific practice. The characterization of a theory into its successes, failures and lacunae is given in terms of the logical conditions that the triple H, C, E should observe. It does not specify, for example, how a lacuna of a theory is identified, that is, how it is possible to determine that there are no initial conditions to give an account of a given evidence as a case of success or failure. Also missing is an explicit way to improve a theory; that is, a way to revise a theory into a better one. We only get the conditions under which two – already given and evaluated – theories are compared with respect to their successes and failures. Kuipers is well aware of the need to introduce the issues of theory revision, as presented in what follows.

Instrumentalist Abduction

In a recent proposal [Kui99], Kuipers uses the above machinery to define the task of *instrumentalist abduction*, namely theory revision aiming at empirical progress. The point of departure is the *evaluation report* of a theory (its successes, failures and lacunae), and the task consists of the following:

Instrumentalist Abduction Task

Search for a revision H_2 of a theory H_1 such that:

H_2 is more successful than H_1, relative to the available data[5].

Kuipers sets an invitation for abduction aiming at empirical progress, namely the design of instrumentalist abduction along symmetric or asymmetric lines. More specific, he proposes the following challenge: *"to operationalize in tableau terms the possibility that one theory is 'more compatible' with the data than another"* [Kui99, p.320]. In this chapter we take up this challenge in the following way. We propose to operationalize the above instrumentalist task for the particular case of successes and lacunae (his Task I). We cast the proposal in our extended framework of semantic tableaux (cf. chapter 4).

It will be clear that while our proposal remains basically the same as that of Kuipers', our representation allows for a more dynamic view of theory evaluation and theory improvement by providing ways to compute appropriate initial conditions for potential successes, and for determining the conditions under which an observation is a case of lacuna, both leading to procedures for theory improvement. However, this chapter does not give all the details of these processes, it only provides the general ideas and necessary proofs.

3. Empirical Progress in (Abductive) Semantic Tableaux

Lacuna Evidence Type Characterization

In what follows we aim to characterize lacuna evidence in terms of the extension type it effects over a certain tableau for a theory when it is expanded with it. The main motivation for this characterization is to answer the question raised in the previous section: can we know in general when a certain formula E qualifies as a case of lacuna with respect to a theory H?

Let me begin by giving a translation of lacuna evidence (cf. subsection 2.2) into tableaux, denoted here as lacuna* in order to distinguish it from Kuipers' original usage:

LACUNA*

Given a tableau $\mathcal{T}(H)$ for a theory, evidence E is a lacuna* of a theory H, when the following conditions hold for all available initial conditions C:

$\mathcal{T}(H) \cup \{\neg E\} \cup \{C\}$ is an open tableau

[5] In the cited paper this is defined as the first part of Task III (Task III.1). Task III.2 requires a further evaluation task and consists of the following: "H_2 remains more successful than H_1, relative to all future data". Moreover, there are two previous tasks: Task I, Task II, which are special cases of Task III, the former amounts to the case of a 'surprising observation' (a potential success or a lacuna) and the task is to expand the theory with some hypothesis (or initial condition) such that the observation becomes a success. This is called *novelty guided abduction*. The latter amounts to the case of an 'anomalous observation' (a failure), and the task is to revise the theory into another one which is able to entail the observation. This one is called *anomaly guided abduction*. In my own terminology of abduction (cf. chapter 2) I refer to the former as *abductive novelty* and to the latter as *abductive anomaly*.

$\mathcal{T}(H) \cup \{E\} \cup \{C\}$ is an open tableau
In this case, neither E confirms nor falsifies H.

Although this is a straightforward translation, it does not capture the precise type of extension that evidence E (and its negation) effects on the tableau. To this end, we recall our proposed distinction between "open", "closed" and "semiclosed" extensions. While the first two are characterized by those formulae which when added to the tableau do not close any open branch, or close all of them respectively, the last one are exemplified by those formulae which close some but not all of the open branches. (For the formal details of the extension characterization in terms of semantic tableaux see chapter 4).

As we are about to prove that, when a formula E (and its negation) effects an open extension over a tableau for a certain theory H, it qualifies as a lacuna*. More precisely:

LACUNA* CLAIM
Given a theory H and evidence E:
IF $\mathcal{T}(H) + \{\neg E\}$ is an open extension and $\mathcal{T}(H) + \{E\}$ is also an open extension,
THEN E is a lacuna* of H.

Proof

(i) To be proved: $\mathcal{T}\ (H) \cup \{\neg E\} \cup \{C\}$ is an open tableau for all C.

(ii) To be proved: $\mathcal{T}(H) \cup \{E\} \cup \{C\}$ is an open tableau for all C.

- **(i)** Let $\mathcal{T}(H) + \{\neg E\}$ be an open extension ($H \not\models E$).
- Suppose there is a C ($C \neq E$ is warranted by the second assumption of conditional confirmation), such that $\mathcal{T}(H) + \{\neg E\} + \{C\}$ is a closed extension.

  Then $\mathcal{T}(H) + \{C\}$ must be a closed extension, that is, $H \models \neg C$, but this contradicts the second assumption for logical independence.
- Therefore,

  $\mathcal{T}(H) + \{\neg E\} + \{C\}$ is not a closed extension, that is, $H, C \not\models E$, thus concluding that $\mathcal{T}(H) \cup \{\neg E\} \cup \{C\}$ is an open tableau.

- **(ii)** Let $\mathcal{T}(H) + \{E\}$ be an open extension ($H \not\models \neg E$).
- Suppose there is a C ($C \neq \neg E$ is warranted by the second assumption of conditional falsification), such that $\mathcal{T}(H) + \{E\} + \{C\}$ is a closed extension.

  Then $\mathcal{T}(H) + \{C\}$ must be a closed extension, that is, $H \models \neg C$, but this contradicts the second assumption of logical independence.

- Therefore,

  $\mathcal{T}(H) + \{E\} + \{C\}$ is not a closed extension, that is, $H, C \not\models \neg E$, thus concluding that $\mathcal{T}(H) \cup \{E\} \cup \{C\}$ is an open tableau.

Notice that in proving the above claim we used two assumptions that have not been stated explicitly for lacunae; one of conditional confirmation in (i) ($C \not\models E$) and the other one of conditional falsification in (ii) ($C \not\models \neg E$). Therefore, we claim these assumptions should also be included in Kuipers' additional logical assumptions as restrictions on lacunae. This fact gives us one more reason to highlight the importance of lacunae as an independent and important case for theory evaluation.

Finally, our characterization of lacuna** evidence type in terms of our proposed tableaux extensions is as follows:

LACUNA**

Given a tableau for a theory $\mathcal{T}(H)$, evidence E is a lacuna** of a theory H whenever:

$\mathcal{T}(H) + \{\neg E\}$ is an open extension

$\mathcal{T}(H) + \{E\}$ is an open extension

In this case, E neither confirms nor falsifies H.

This account allows one to characterize lacuna type evidence in terms of theory and evidence alone, without having to consider any potential initial conditions. This result is therefore attractive from a computational perspective, since it prevents the search of appropriate initial conditions when there are none.

The proof of the above lacuna claim assures that our characterization implies that of Kuipers', that is, open extensions imply open tableaux (lacuna** implies lacuna*), showing that our characterization of lacuna** is stronger than the original one of lacuna*. However, the reverse implication (lacuna* implies lacuna**) is not valid. An open tableau for $\mathcal{T}(H) \cup \{\neg E\} \cup \{C\}$ need not imply that $\mathcal{T}(H) + \{\neg E\}$ is an open extension. Here is a counterexample:

Let $H = \{a \rightarrow b, a \rightarrow d, b \rightarrow c\}$

$E = \{b\}$ and $C = \{\neg d\}$

Notice that $\mathcal{T}(H) \cup \{\neg b\} \cup \{\neg d\}$ is an open tableau, but both $\mathcal{T}(H) + \{\neg b\}$ and $\mathcal{T}(H) + \{\neg d\}$ are semiclosed extensions.

Examples

In what follows we show the fundamentals of our approach in the framework of Semantic Tableaux (cf. chapter 4) , via an example on a lacuna-type evidence followed by one of theory improvement.

Lacuna

Let H_1 be a theory and E an evidence. $H_1 = \{a \rightarrow b\}, E = \{c\}$

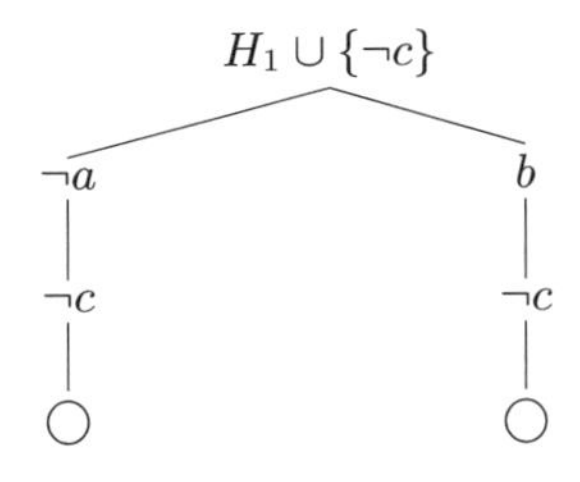

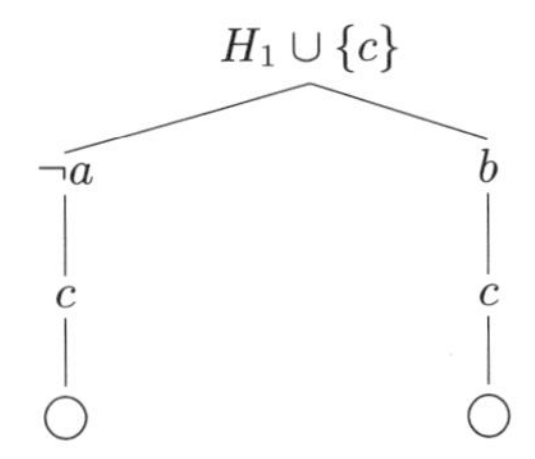

In this case, neither evidence (c) nor its negation ($\neg c$) are valid consequences of the theory H_1, as shown by the two open tableaux above. Moreover, it is not possible to produce any (literal) formula acting as initial condition in order to make the evidence a case of success or failure (unless the trivial case is considered, that is, to add c or $\neg c$). For this particular example it is perhaps easy to be aware of this fact (a and $\neg b$ are the only possibilities and neither one closes the whole tableau), but the central question to tackle is, in general, how can we know when a certain formula qualifies as a case of lacuna with respect to a certain theory?

We are then faced with a genuine case of a lacuna, for which the theory holds no opinion about the given evidence, either affirmative or negative: it simply cannot account for it.

As we saw previously, we characterize formulae as lacunae of a theory whenever neither one of them nor their negations close any open branch when the tableau is extended with them. As we have pointed out, this kind of evidence not only shows that the theory is insufficient to account for it, but also suggests a condition for theory improvement, thus indicating the possibility of empirical progress leading to a procedure to perform the instrumentalist task (cf. [Kui99]).

Theory Improvement

In what follows we give an example of the extension of theory H_1 into a theory H_2, differing in that formula c is a lacuna in the former and a success in the latter. One way to expand the tableau is by adding the formula $a \rightarrow c$ to H_1 obtaining H_2 as follows:

Let $H_2 = \{a \rightarrow b, a \rightarrow c\}, E = \{c\}$

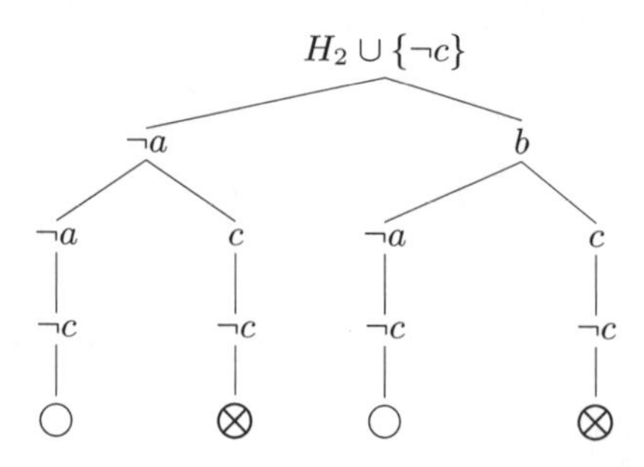

This tableau has still two open branches, but the addition of a new formula to H_1 (resulting in H_2), converts evidence c into a candidate for success or failure, since it has closed an open branch (in fact two), and thus its extension qualifies as *semi–closed*. It is now possible to calculate an appropriate initial condition to make c a success in H_2, namely a. Thus, c is a success of H_2 with respect to a.

But H_1 may be extended by means of other formulae, possibly producing even *better* theories than H_2. For instance, by extending the original tableau instead with $b \rightarrow c$ the following is obtained:

Let $H_3 = \{a \rightarrow b, b \rightarrow c\}, E = \{c\}$

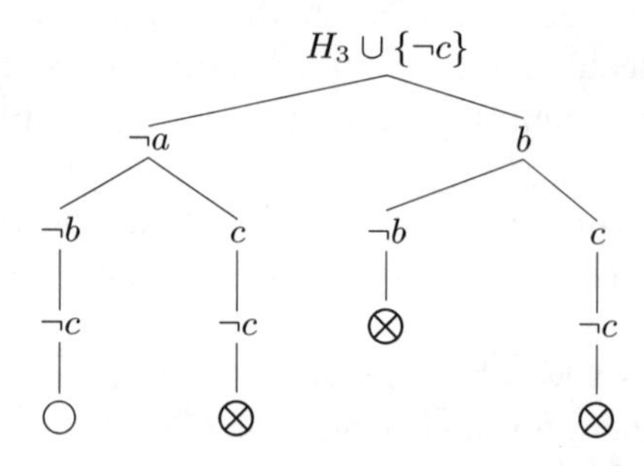

As before, c is a success with respect to initial condition a, but in addition, c is also a success with respect to initial condition b. Formula a as well as b close the open branch in a non-trivial way. Therefore, in H_3 we gained one success over H_2. Shall we then consider H_3 a better theory than both H_1 and H_2?

According to Kuipers' criterion for theory comparison (cf. section 2.3), both H_2 and H_3 are more successful theories than H_1, since the subset of successes in H_1 is a proper subset of both the set of successes in H_2 and of H_3. Moreover, H_3 is more successful than H_2, since in the former case c is a success with respect to two initial conditions and only with respect to one in the latter case.

We leave here our informal presentation of the extended method of semantic tableaux applied to empirical progress leading to the instrumentalist task, as conceived by Kuipers.

4. Discussion and Conclusions

Success and Failure Characterizacion?

The reader may wonder whether it is also possible to characterize successes and failures in terms of their extension types. Our focus in this chapter has been on lacuna, so we limit ourselves to presenting the results we have found so far, omitting their proofs:

Success IF E is a Success of H THEN

$\mathcal{T}(H) + \{\neg E\}$ is a semiclosed extension and there is an initial condition available C for which $\mathcal{T}(H) + \{C\}$ is a semiclosed extension.

Failure IF E is a failure of H THEN

$\mathcal{T}(H) + \{E\}$ is a semiclosed extension and there is an initial condition C available for which $\mathcal{T}(H) + \{C\}$ is a semiclosed extension.

For these cases it turns out that the corresponding valid implication is the reverse of the one with respect to that of lacunae. This means that closed tableaux imply semiclosed extensions, but not the other way around. In fact, the above counterexample[6] is also useful to illustrate this failure: both $\mathcal{T}(H)+\{\neg b\}$ and $\mathcal{T}(H)+\{\neg d\}$ are semiclosed extensions, but E is not a success of H, since $\mathcal{T}(H) \cup \{\neg b\} \cup \{\neg d\}$ is an open tableau.

Conclusions

The main goal of this chapter has been to address Kuipers' challenge in [Kui99], namely to operationalize the task of *instrumentalist abduction*. In particular, the central question raised here concerns the role of lacunae in the dynamics of empirical progress, both in theory evaluation and in theory improvement. That is, the relevance of lacunae for the success and the improvement questions.

Kuipers' approach to empirical progress does take account of lacunae for theory improvement, by including them in the evaluation report of a theory (together with its successes and failures), but it fails to provide a full characterization of them, in the same way it does for successes and failures (the former via the notion of confirmation, and the latter via the notion of falsification). Moreover, this approach does not specify precisely how a lacuna of a theory is

[6] $H = \{a \to b, a \to d, b \to c\}$ and $E = \{b\}$.

recognized, that is, how it is possible to identify the absence of initial conditions which otherwise give an account of certain evidence as a case of success or failure.

Our reformulation of Kuipers' account of empirical progress in the framework of (extended) semantic tableaux is not just a matter of a translation into a mathematical framework. There are at least two reasons for this claim. On the one hand, with tableaux it is possible to determine the conditions under which a phenomenon is a case of lacuna for a certain theory, and on the other hand, this characterization leads to a procedure for theory improvement. Regarding the first issue, we characterize formulae as lacunae of a theory whenever neither one of them nor their negations close any open branch (when the tableau is extended with them), they are *open extensions* in our terminology, and we prove that our formal characterization of lacuna implies that of Kuipers'. As for the latter assertion, lacunae type evidences not only show that the theory is insufficient to account for them, but also suggests a condition for theory improvement, thus indicating possibilities of empirical progress leading to a procedure for performing the instrumentalist task.

We have suggested that a fresh look into classical methods in logic may be used not only to represent the corroboration of theories with respect to evidence, that is, to answer the truth question, but also to evaluate theories with respect to their evidence as well as to generate better theories than the original ones. To this end, we have presented (though informally) the initial steps to answer the success and improvement questions, showing that given a theory and an evidence, appropriate initial conditions may be computed in order to make that evidence a success (failure) with respect to a theory; and when this is not possible, we have proposed a way to extend a theory and thus improve it in order to account for its lacunae.

Chapter 7

PRAGMATISM

1. Introduction

In this chapter, I present the philosophical doctrine known as pragmatism, as proposed by Charles Peirce, namely, as a method of reflexion with the aim at clarifying ideas and guided at all moments by the ends of the ideas it analyzes. Pragmatism puts forward an epistemic aim with an experimental solution, and does so by following the *pragmatic maxim*, the underlying precept to fix a belief, and accordingly produce its corresponding habits of action.

As it turns out, abduction has close connections to pragmatism, the former is indeed the basis for the latter. This connection is however, a complex one. On the one hand, to link abduction with a method of reflection, we must conceive it as an epistemic process for logical inquiry of some kind, which transforms the *surprising phenomenon* into a *settled belief.* Furthermore, pragmatism is centered in the *pragmatic maxim*, a guide to calculate practical effects of conceptions. Thus, on the other hand, in order to attend the experimental aspect of pragmatism in its connection to abduction, we must attend further aspects of an abductive hypothesis, and these necessarily go beyond its logical formulation.

This chapter is naturally divided into four parts. After this introduction, in the second part (section 2), we present Peirce's notion of pragmatism, its motivation and purposes, and the pragmatic maxim. In the third part (section 3), we present Peirce's development of the notion of abduction as well as his epistemic view. The abductive formulation can thus be seen as a process of belief acquisition by which a surprising fact generates a doubt that is appeased by a belief, an abductive explanation. In the fourth part (section 4), we present Peirce's notion of pragmatism revisited, that is, in connection with his epistemology and his (later) view on abduction. Pragmatism is a method of reflexion, and that constitutes its epistemic aspect, but it is also intimately linked with action, with those habits of conduct involved in a belief, which is what constitutes the experimental aspect that gives meaning to things. In the fifth and final part of this chapter (section 5), we put forward our conclusions relating aspects of Peirce's pragmatism to abduction and epistemology.

2. Pragmatism

Pragmatism and Pragmaticism

William James reports it was Charles Peirce who engendered the philosophical doctrine known as *Pragmatism.* Just as other notions coined by Peirce, the choice of this term was carefully selected, this time with the intention of capturing both sides of this concept: an epistemic as well as a practical one:

> *'Now quite the most striking feature of the new theory was its recognition of an inseparable connection between rational cognition and rational purpose; and that consideration it was which determined the preference for the name* **pragmatism***'.* [CP, 5.412]

The term itself was taken from Kant's 'pragmatisch', meaning 'in relation to some (well) defined human purpose' and aims to be distinguished from the other Kantian term 'praktisch', which rather belongs to a region of thought *'where no mind of the experimentalist type can ever make sure of solid ground under his feet.'* [CP, 5.412]

Later on, however, Peirce preferred to call it *pragmaticism*, this time to mark a difference with the notion of pragmatism proposed by W. James. The history of this term is fairly well described by Peirce in the following passage [CP, 6.882]:

> *"In 1871, in a Metaphysical Club in Cambridge, Massachusetts, I used to preach this principle as a sort of logical gospel, representing the unformulated method followed by Berkeley, and in conversation about it I called it "Pragmatism". In December [November] 1877 and January 1878 I set forth the doctrine in the Popular Science Monthly; and the two parts of my essay were printed in French in the Revue Philosophique, volumes vi and vii. Of course, the doctrine attracted no particular attention, for, as I had remarked in my opening sentence, very few people care for logic. But in 1897 Professor James remodelled the matter, and transmogrified it into a doctrine of philosophy, some parts of which I highly approved, while other and more prominent parts I regarded, and still regard, as opposed to sound logic. About the time Professor Papini discovered, to the delight of the Pragmatist school, that this doctrine was incapable of definition, which would certainly seem to distinguish it from every other doctrine in whatever branch of science, I was coming to the conclusion that my poor little maxim should be called by another name; and accordingly, in April, 1905 I renamed it* **Pragmaticism***"*[1].

The doctrine of pragmatism is proposed as a philosophical method rather than a mere philosophy of life, thereby marking yet another difference with James' pragmatism:

[1]Despite this clarification, for the aims of this chapter, we refer to the Peircean doctrine as pragmatism.

> *"It will be seen [from the original statement] that pragmatism is not a Weltanschauung (worldview) but is a method of reflexion having for its purpose to render ideas clear".* [CP, 5.13].

Pragmatism and Meaning

Pragmatism is a methodological principle responding to questions concerning a theory of meaning; it does not pretend to define truth or reality, but rather to determine the meaning of terms or propositions:

> *"Suffice it to say once more that pragmatism is, in itself, no doctrine of metaphysics, no attempt to determine any truth of things. It is merely a method of ascertaining the meanings of hard words and of abstract concepts... All pragmatists will further agree that their method of ascertaining the meanings of words and concepts is no other than that experimental method by which all the successful sciences (in which number nobody in his senses would include metaphysics) have reached the degrees of certainty that are severally proper to them today; this experimental method being itself nothing but a particular application of an older logical rule, "By their fruits ye shall know them."* [CP, 5.464, 5.465]

The notion of 'the meaning of a concept' in this pragmatic terms (cf. [CP, 6.481]) is that it is acquired through the following conditions by which a master of its use is attained. In the first place, it is required to learn to recognize a concept in whatever of its manifestations, and this is achieved by an extensive familiarization with its instances. In the second place, it is required to carry out an abstract logical analysis of the concept, getting to the bottom of its elemental constitutive parts. But these two requirements are not yet sufficient to know the nature of a concept in its totality. For this, it is in addition necessary to discover and recognize those habits of conduct that the belief in the truth of the concept in question naturally generates, that is, those habits which result in a sufficient condition for the truth of a concept in any theme or imaginable circumstance.

The Pragmatic Maxim

Pragmatism is centered in its *pragmatic maxim*, the underlying guiding principle of this doctrine. In its in its original formulation, this maxim reads as follows:

> *"Consider what effects that might conceivably have practical bearing you conceive the object of your conception to have. Then your conception of those effects is the whole of your conception of the object".* (Reveu philosophique VII, [CP, 5.18]).

The core of this principle is that the conception of an object constitutes its conceivable practical effects, manifest in habits of action, as the following quote suggests:

> *'To develop its meaning, we have, therefore, simply to determine what habits it produces, for what a thing means is simply what habits it involves. Now, the identity of a habit depends on how it might lead us to act, not merely under such circumstances as are*

likely to arise, but under such as might possibly occur, no matter how improbable they may be'. [CP, 5.400]

Peirce adds the following in regard to the notion of 'conduct':

"It is necessary to understand the word 'conduct', here, in the broadest sense. If, for example, the predication of a given concept were to lead to our admitting that a given form of reasoning concerning the subject of which it was affirmed was valid, when it would not otherwise be valid, the recognition of that effect in our reasoning would decidedly be a habit of conduct". [CP, 6.481].

"For the maxim of pragmatism is that a conception can have no logical effect or import differing from that of a second conception except so far as, taken in connection with other conceptions and intentions, it might conceivably modify our practical conduct differently from that second conception". [CP, 5.196].

The method proposed by this maxim provides a regulative principle for the evaluation of beliefs and serves as a guide for our actions, and in this respect it is a normative principle. The pragmatic maxim becomes the rule to achieve the main aim of the whole doctrine, namely to clarify ideas.

3. Abduction and Epistemology

Abduction in Peirce

The notion of abduction in the work of Charles Peirce is entangled in many aspects of his philosophy and is therefore not limited to the conception of abduction as a logical inference of its own. The notions of logical inference and of validity that Peirce puts forward go beyond our present understanding of what logic is about. On the one hand, they are linked to his epistemology, a dynamic view of thought as logical inquiry, and correspond to a deep philosophical concern, that of studying the nature of synthetic reasoning.

The intellectual enterprise of Charles Sanders Peirce, in its broadest sense, was to develop a semiotic theory, in order to provide a framework to give an account for thought and language. With regard to our purposes, the fundamental question Peirce addressed was how synthetic reasoning was possible[2]. Very much influenced by the philosophy of Immanuel Kant, Peirce's aim was to extend his categories and correct his logic:

"According to Kant, the central question of philosophy is 'How are synthetical judgments a priori possible?' But antecedently to this comes the question how synthetical judgments in general, and still more generally, how synthetical reasoning is possible at all. When the answer to the general problem has been obtained, the particular one will be comparatively simple. This is the lock upon the door of philosophy". ([CP, 5.348], quoted in [Hoo92], page 18).

[2] Cf. [FK00] for Peirce's classification of inferences into analytic and synthetic.

Peirce proposes abduction to be the logic for synthetic reasoning, that is, a method to acquire new ideas. As already noted in chapter 2, he was the first philosopher to give to abduction a logical form.

The development of a logic of inquiry occupied Peirce's thought since the beginning of his work. In the early years he thought of a logic composed of three modes of reasoning: deduction, induction and hypothesis each of which corresponds to a syllogistic form, illustrated by the following, often quoted example [CP, 2.623]:

DEDUCTION

Rule.– All the beans from this bag are white.
Case.– These beans are from this bag.
Result.– These beans are white.

INDUCTION

Case.– These beans are from this bag.
Result.– These beans are white.
Rule.– All the beans from this bag are white.

HYPOTHESIS

Rule.– All the beans from this bag are white.
Result.– These beans are white.
Case.– These beans are from this bag.

Of these, deduction is the only reasoning which is completely certain, inferring its 'Result' as a necessary conclusion. Induction produces a 'Rule' validated only in the 'long run' [CP, 5.170], and hypothesis merely suggests that something may be 'the Case' [CP, 5.171]. The evolution of his theory is also reflected in the varied terminology he used to refer to abduction; beginning with *presumption* and *hypothesis* [CP, 2.776,2.623], then using *abduction* and *retroduction* interchangeably [CP, 1.68,2.776,7.97].

Later on, Peirce proposed these types of reasoning as the stages composing a method for logical inquiry, of which abduction is the beginning:

> *"From its [abductive] suggestion deduction can draw a prediction which can be tested by induction."* [CP, 5.171].

Abduction plays a role in direct perceptual judgments, in which:

> *"The abductive suggestion comes to us as a flash"* [CP, 5.181].

As well as in the general process of invention:

> *"It [abduction] is the only logical operation which introduces any new ideas"* [CP, 5.171].

In all this, abduction is both *"an act of insight and an inference"* as has been suggested in [And86]. These explications do not fix one unique notion. Peirce refined his views on abduction throughout his work. He first identified abduction with the syllogistic form above, to later enrich this idea by the more general conception of:

"the process of forming an explanatory hypothesis" [CP, 5.171].

And also referring to it as:

"The process of choosing a hypothesis" [CP, 7.219].

Something that suggests that he did not always distinguish clearly between the construction and the selection of a hypothesis. In any case, this later view gives place to the logical formulation of abduction ([CP, 5.189]), which we reproduced again (cf. chapter 2) as follows:

The surprising fact, C, is observed.

But if A were true, C would be a matter of course.

Hence, there is reason to suspect that A is true.

We recall from chapter 2 that in addition to this formulation which makes up the explanatory constituent of abductive inference, there are two other aspects to consider for an explanatory hypothesis, namely its being testable and economic. While the former sets up a requirement in order to give an empirical account of the facts, the latter is a response to the practical problem of having innumerable hypotheses to test and points to the need of having a criterion to select the best explanation amongst the testable ones. We will return to the former additional criterion later in the chapter when the connection to pragmatism is put forward.

Interpreting Peirce's Abduction

The notion of abduction has puzzled Peirce scholars all along. Some have concluded that Peirce held no coherent view on abduction at all [Fra58], others have tried to give a joint account with induction [Rei70] and still others claim it is a form of inverted modus ponens [And86]. A more modern view is found in [Kap90] who interprets Peirce's abduction as a form of heuristics. An account that tries to make sense of the two extremes of abduction, both as a guessing instinct and as a rational activity is found in [Ayi74]. This last approach continues to present day. While [Deb97] proposes to reinterpret the concept of rationality to account for these two aspects, [Gor97] shows abductive inference in language translation, a process in which the best possible hypothesis is sought using instinctive as well as rational elements of translation. Thus, abductive inference is found in a variety of contexts. To explain abduction in perception, [Roe97] offers a reinterpretation of Peirce's abductive formulation, whereas [Wir97] uses the notion of 'abductive competence' to account for language interpretation.

This diversity suggests that Peirce recognized not only different types of reasoning, but also several degrees within each one, and even merges between the types. In the context of perception he writes:

"The perceptual judgements, are to be regarded as extreme cases of abductive inferences" [CP, 5.181].

Abductory induction, on the other hand, is suggested when some kind of guess work is involved in the reasoning [CP, 6.526][3]. Anderson [And87] also recognizes several degrees in Peirce's notion of creativity. A nice concise account of the development of abduction in Peirce, which clearly distinguishes three stages in the evolution of his thought is given in [Fan70]. Another key reference on Peirce's abduction, in its relation to creativity in art and science is found in [And87][4].

Peirce's Epistemology

In Peirce's epistemology, thought is a dynamic process, essentially an interaction between two states of mind: doubt and belief. While the essence of the latter is the *"establishment of a habit which determines our actions"* [CP, 5.388], with the quality of being a calm and satisfactory state in which all humans would like to stay, the former *"stimulates us to inquiry until it is destroyed"* [CP, 5.373], and it is characterized by being a stormy and unpleasant state from which every human struggles to be freed:

"The irritation of doubt causes a struggle to attain a state of belief". [CP, 5.374].

Peirce speaks of a state of belief and not of knowledge. Thus, the pair 'doubt-belief' is a cycle between two opposite states. While belief is a habit, doubt is its privation. Doubt, however, is not a state generated at will by raising a question, just as a sentence does not become interrogative by putting a special mark on it, there must be a *real* and *genuine* doubt:

"Genuine doubt always has an external origin, usually from surprise*; and that it is as impossible for a man to create in himself a genuine doubt by such an act of the will as would suffice to imagine the condition of a mathematical theorem, as it would be for him to give himself a genuine surprise by a simple act of the will".* ([CP, 5.443]).

Moreover, it is surprise what breaks a habit:

"For belief, while it lasts, is a strong habit, and as such, forces the man to believe until some **surprise** *breaks up the habit".* ([CP, 5.524], my emphasis).

And Peirce distinguishes two ways to break a habit:

"The breaking of a belief can only be due to some novel experience*" [CP, 5.524] or "...until we find ourselves confronted with some experience* **contrary to those expectations**." ([CP, 7.36], my emphasis).

[3] Peirce further distinguishes three kinds of induction [CP, 2.775,7.208], and even two kinds of deduction [CP, 7.224].

[4] This multiplicity returns in artificial intelligence. [Fla96b] suggests that some confusions in modern accounts of abduction in AI can be traced back to Peirce's two theories of abduction: the earlier syllogistic one and the later inferential one. As to more general semiotic aspects of Peirce's philosophy, another proposal for characterizing abduction in AI is found in [Kru95].

Peirce's epistemic model proposes two varieties of surprise as the triggers for every inquiry, which we relate to the previously proposed novelty and anomaly (cf. chapter 2). Moreover, these are naturally related to the epistemic operations for belief change in AI (cf. [Ali00b] and chapter 8).

4. Pragmatism Revisited

A method of inquiry aiming to clarify ideas, by giving meanings to objects according to those habits of action involved, is what makes pragmatism an active philosophical method of reflexion. Pragmatism puts forward an epistemic aim with an experimental solution, precisely providing an *'unseparable connection between rational cognition and rational purpose'* [CP, 5.412].

The nature of this method is logical and not metaphysical: *"I make pragmatism to be a mere maxim of logic instead of a sublime principle of speculative philosophy"* [CP, 7.220]. Moreover, the method itself is guided by 'the pragmatic maxim', the underlying principle by which a belief is fixed; it is a rule to achieve the highest degree of clarity in the apprehension of ideas. These are the motivations and aims of pragmatism; being abduction the underlying logic of this process: *"if you carefully consider the question of pragmatism you will see that it is nothing else than the question of the logic of abduction"* [CP, 4.196].

Abduction is then proposed as an epistemic notion as well as a pragmatic one. How can this be possible? As previously mentioned (cf. chapter 2) for a hypothesis to be considered an abductive explanation it should be *explanatory*, *testable* and *economic*. Abduction has indeed a logical formulation, which shapes its explanatory aspect, characterizing the argumentative format that all candidates for abductive hypotheses should observe. It is the first of these aspects what is related to his epistemology. In particular, our previous analysis was centered on the role played by the element of *surprise* in this formulation; and on its connection to the epistemic transition between the states of doubt and belief. The second of these aspects is what is relevant to the connection to pragmatism, for this doctrine provides a maxim which precisely characterizes what is to count as an explanatory hypothesis based on its being subject to experimental verification.

Therefore, in order to understand the relationship between abduction and pragmatism we must go beyond the logical form of abduction in two respects. On the one hand, abduction must be understood as an epistemic process for the acquisistion of belief and on the other hand, we must take into consideration the experimental requirement. In what follows we will present the connections between abduction and epistemology, followed by the connections between each of these with pragmatism.

Abduction and Epistemology

The connection between abductive logic and the epistemic transition between the mental states of doubt and belief is clearly seen in the fact that the surprise is both the trigger of abductive reasoning –as indicated by the first premise of the logical formulation of abduction– as well as that of the doubt state when an belief habit has been broken.

The overall cognitive process showing abductive inference as an epistemic process can be depicted as follows: a novel or an anomalous experience (cf. chapter 2) gives place to a surprising phenomenon, generating an state of doubt which breaks up a belief habit and accordingly triggers abductive reasoning. The goal of this type of reasoning is precisely to explain the surprising fact and therefore *soothe* the state of doubt. It is 'soothe' rather than 'destroy' for an abductive hypothesis has to be put into test and be economic, attending to the further criteria Peirce proposed. The abductive explanation is simply a suggestion that has to be put to test before converting itself into a belief.

Epistemology and Pragmatism

Two texts concerned with inquiry mark the first places in which Peirce presents the ideas which shape the doctrine of pragmatism, namely 'The fixation of belief' (1877, [CP, 5.358–387]) and 'How to make our ideas clear' (1878, [CP, 5.388–5.410]). The first of these texts, presents the end of inquiry as the settlement of opinion as well as the methods for fixing a belief. It worships a logical method over the methods based on an appeal to authority, on tenacity and on the a priori: *'the very first lesson that we have a right to demand that logic shall teach us is, how to make our ideas clear'* [CP, 5.393]. The primary function of thought is to produce a belief:

> *'The action of thought is excited by the irritation of doubt, and ceases when belief is attained; so that the production of belief is the sole function of thought'*.[CP, 5.394].

The second text reiterates the importance of a method of a logical kind and introduces the pragmatic maxim, as the guiding principle to achieve the highest degree of clarity in the apprehension of ideas: *'Our idea of anything is our idea of its sensible effects'*[CP, 5.401]. The effects of our conceptions must be produced in order to develop their meaning, and they become alive through habits of action. In fact, a late definition of pragmatism given by Peirce, reads as follows:

> *'What the true definition of Pragmatism may be, I find it very hard to say; but in my nature it is a sort of instinctive attraction for* **living facts***'*.[CP, 5.64]

Abduction and Pragmatism

In regard to the connection between abduction and pragmatism, we find in Peirce's writings notes for a conference "Pragmatism – Lecture VII–" (of which there is evidence it was never delivered). This conference is composed by four sections[5], of which only the third one, "Pragmatism – The Logic of Abduction" (cf. [CP, 5.195] – [CP, 5.206]) becomes relevant to our discussion. In this section of the conference, Peirce states that the question of pragmatism is nothing else than the logic of abduction, as suggested by the following quote:

> *"Admitting, then, that the question of Pragmatism is the question of Abduction, let us consider it under that form. What is good abduction? What should an explanatory hypothesis be to be worthy to rank as a hypothesis? Of course, it must explain the facts. But what other conditions ought it to fulfill to be good? The question of the goodness of anything is whether that thing fulfills its end. What, then, is the end of an explanatory hypothesis? Its end is, through subjection to the test of experiment, to lead to the avoidance of all surprise and to the establishment of a habit of positive expectation that shall not be disappointed. Any hypothesis, therefore, may be admissible, in the absence of any special reasons to the contrary, provided it be capable of experimental verification, and only insofar as it is capable of such verification. This is approximately the doctrine of pragmatism. But just here a broad question opens out before us. What are we to understand by experimental verification? The answer to that involves the whole logic of induction".* [CP, 5.198].

Peirce puts forward the pragmatic method as that providing a maxim that completely characterizes the admissibility of explanatory hypotheses. On the one hand, it is required that every hypothesis is subject to experimental corroboration (verification in Peircean terms) and on the other hand, this corroboration is manifest in habits of conduct which eliminate the surprise of doubt in question.

Another aspect of equal importance, but apparently outside the boundaries of the logic of abduction, is concern with the very notion of experimental corroboration, which according to the preceding text, Peirce poses in the logic of induction. Moreover, this aspect may go beyond empirical experimentation:

> *"If pragmatism is the doctrine that every conception is a conception of conceivable practical effects, it makes conception reach far beyond the practical. It allows for any flight of imagination, provided this imagination ultimately alights upon a possible practical effect; and thus many hypothesis may seem at first glance to be excluded by the pragmatical maxim that are really not excluded".* [CP, 5.196].

Besides the suggestion in this quote that experimental corroboration does not limit itself to merely empirical experimentation, for it gives place to thought experiments and other manifestations in the realm of ideas, and it also suggests that the process of conceiving the possible effects of a certain conception already involves the mere action of experimentation, for in the calculation of the practical effects manifest in the habits of conduct that a certain explanatory hypothesis produces, we are already in the territory of experimental corroboration, being these found in the practical realm or in the world of ideas.

[5]The section titles are the following ones: "The Three Cotary Propositions", "Abduction and Perceptual Judgments", "Pragmatism – the Logic of Abduction" and "The Two Functions of Pragmatism".

5. Discussion and Conclusions

In this chapter we analyzed the doctrine of pragmatism in the work of Charles Peirce, as a philosophical method of reflexion guided by the pragmatic maxim, the underlying principle by which a belief is fixed, generating habits of action. The outcome of our analysis shows that pragmatism is closely connected with Peirce's abduction. A direct link is found in the requirement regarding experimental corroboration for explanatory hypotheses, those which comply with the logical formulation of abduction. The experimental corroboration requirement raises the significance of the pragmatic dimension of the logic of abduction and gives an answer to the hypotheses selection problem of those that are explanatory. For Peirce, the experimental corroboration of explanatory hypotheses goes beyond verification, as it requires of a calculation of its possible consequences of effects; those that produce new habits of conduct, being these epistemic or practical.

A natural consequence of this analysis for abduction, is that the interpretation of Peirce's formulation goes beyond that of a logical argument to become an epistemic process which may be described as follows: a novel or an anomalous fact gives place to a surprising fact, generating a state of doubt. Therefore, abductive reasoning is triggered, which consists on explaining the surprising fact by providing an explanatory hypothesis that plays the role of a new belief. And this state will remain as such until another fact is encountered, thus continuing the epistemic doubt–belief cycle.

It is appealing that this view has close connections with theories of belief revision in artificial intelligence[6]. It also involves extending the traditional view of abduction in AI, to include cases in which the observation is in conflict with the theory, as we have suggested in our taxonomy (cf. chapter 2.).

[6]In the case of an abductive novelty, the explanation is assimilated into the theory by the operation of expansion. In the case of an abductive anomaly, the operation of revision is needed to modify the theory and incorporate the explanation. Cf. chapter 8 for a detailed analysis of this connection.

Discussion and Conclusions

Chapter 8

EPISTEMIC CHANGE

1. Introduction

Notions related to explanation have also emerged in theories of belief change in AI. One does not just want to incorporate new beliefs, but often also, to justify them. The main motivation of these theories is to develop logical and computational mechanisms to incorporate new information to a scientific theory, data base or set of beliefs. Different types of change are appropriate in different situations. Indeed, the pioneering work of Carlos Alchourrón, Peter Gärdenfors and David Makinson (often referred as the AGM approach) [AGM85], proposes a normative theory of epistemic change characterized by the conditions that a rational belief change operator should satisfy.

Our discussion of epistemic change is in the same spirit, taking a number of cues from their analysis. We will concentrate on *belief revision*, where changes occur only in the theory. The situation or world to be modelled is supposed to be static, only new information is coming in. (The other type of epistemic change in AI which accounts for a changing world is called *update*, which we will briefly discuss at the end of the chapter.)

This chapter is naturally divided into four parts. After this introduction, in which we give a brief introduction of abduction as a process of belief revision, in the second part (section 2) we review theories of belief revision in AI and then propose abduction as belief revision, relating our previous 'abductive triggers' (chapter 2) with the epistemic attitudes in the belief revision theories and defining operations for abductive expansion, revision and contraction of their own. Finally, we show that abduction as a theory of epistemic change is both committed to the foundationalist and the coherentist epistemological stances of belief revision theories. In the third part (section 3), we propose an implementation in tableaux of abduction as a process of belief revision. We extend our previous analysis of chapter 4 as to include contraction and revision operations over tableaux. We propose two different strategies for contraction (without committing ourselves to any one of them) and outline procedures for computing explanations. In the fourth and final part of this chapter (section 4), we offer an analysis of previous sections with respect both to the relation of abduction to belief revision and to the algorithmic sketch we presented previously. We then put forward our conclusions and briefly mention some other related work in this direction.

Generally speaking, this chapter shows that the notion of abduction goes beyond that of logical inference and that it may indeed be interpreted as a process for belief revision, with clear connections with the existing literature in this field. It also shows that semantic tableaux may have further uses implementing contraction and revision operations, which go way beyond the standard ones. It also involves extending the traditional view of abduction in AI, to include cases in which the observation is in conflict with the theory, as suggested in our taxonomy in chapter 2.

2. Abduction as Epistemic Change

Theories of Belief revision in AI

We shall expand on the brief introduction given in chapter 2, highlighting aspects that distinguish different theories of belief revision. This sets the scene for approaching abduction as a similar enterprise[1].

The basic elements of this theory are the following. Given a consistent theory Θ closed under logical consequence, called the belief state, and a sentence φ, the incoming belief, there are three *epistemic attitudes* for Θ with respect to φ: either φ is accepted ($\varphi \in \Theta$), φ is rejected ($\neg\varphi \in \Theta$), or φ is undetermined ($\varphi \notin \Theta, \neg\varphi \notin \Theta$). Given these attitudes, the following operations characterize the kind of belief change φ brings into Θ, thereby effecting an epistemic change in the agent's currently held beliefs:

- Expansion

  A new sentence is added to Θ regardless of the consequences of the larger set to be formed. The belief system that results from expanding Θ by a sentence φ together with the logical consequences is denoted by $\Theta + \varphi$.

- Revision

  A new sentence that is (typically) inconsistent with a belief system Θ is added, but in order that the resulting belief system be consistent, some of

[1]The material of this section is mainly based on [Gar92], with some modifications taken from other approaches. In particular, in our discussion, belief revision operations are not required to handle incoming beliefs together with all their logical consequences.

the old sentences in Θ are deleted. The result of revising Θ by a sentence φ is denoted by $\Theta * \varphi$.

- Contraction

  Some sentence in Θ is retracted without adding any new facts. In order to guarantee the deductive closure of the resulting system, some other sentences of Θ may be given up. The result of contracting Θ with respect to sentence φ is denoted by $\Theta - \varphi$.

Of these operations, revision is the most complex one. Indeed the three belief change operations can be reduced into two of them, since revision and contraction may be defined in terms of each other. In particular, revision here is defined as a composition of contraction and expansion: first contract those beliefs of Θ that are in conflict with φ, and then expand the modified theory with sentence φ (known as 'Levi's identity'). While expansion can be uniquely and easily defined ($\Theta + \varphi = \{\alpha \mid \Theta \vee \{\varphi\} \vdash \alpha\}$), this is not so with contraction or revision, as several formulas can be retracted to achieve the desired effect. These operations are intuitively non-deterministic. A simple example (cf. example 3 from chapter 2) to illustrate this point is the following:

Θ: $r, r \rightarrow w$.

φ: $\neg w$.

In order to incorporate φ into Θ and maintain consistency, the theory must be revised. But there are two possibilities for doing this: deleting either of $r \rightarrow w$ or r allows us to then expand the contracted theory with $\neg w$ consistently. Several formulas can be retracted to achieve the desired effect, thus it is impossible to state in purely logical or set-theoretical terms which of these is to be chosen.

Therefore, an additional criterion must be incorporated in order to fix which formula to retract. Here, the general intuition is that changes on the theory should be kept 'minimal', in some sense of informational economy[2].

Moreover, epistemic theories in this tradition observe certain 'integrity constraints', which concern the theory's preservation of consistency, its deductive closure and two criteria for the retraction of beliefs: the loss of information should be kept minimal and the less entrenched beliefs should be removed first.

These are the very basics of the AGM approach. In practice, however, full-fledged systems of belief revision can be quite diverse. They differ in at least three aspects: (a) belief state representation, (b) characterization of the operations of epistemic change (via postulates or constructively), and (c) epistemological stance.

[2]Various ways of dealing with this issue occur in the literature. I mention only that in [Gar88]. It is based on the notion of *entrenchment*, a preferential ordering which lines up the formulas in a belief state according to their importance. Thus, we may retract those formulas that are the 'least entrenched' first. For a more detailed reference as to how this is done exactly, see [GM88].

Regarding the first aspect (a), we find there are essentially three ways in which the background knowledge Θ is represented: (i) belief sets, (ii) belief bases, or (iii) possible world models. A belief set is a set of sentences from a logical language L closed under logical consequence. In this classical approach, expanding or contracting a sentence in a theory is not just a matter of addition and deletion, as the logical consequences of the sentence in question should also be taken into account. The second approach emerged in reaction to the first. It represents the theory Θ as a *base for a belief set* B_Θ, where B_Θ is a finite subset of Θ satisfying $Cons(B_\Theta) = \Theta$. (That is, the set of logical consequences of B_Θ is the classical belief state). The intuition behind this is that some of the agent's beliefs have no independent status, but arise only as inferences from more basic beliefs. Finally, the more semantic approach (iii) moves away from syntactic structure, and represents theories as sets W_Θ of possible worlds (i.e., their models). Various equivalences between these approaches have been established in the literature (cf. [GR95]).

As for the second aspect (b), operations of belief revision can be given either 'constructively' or merely via 'postulates'. The former approach is more appropriate for algorithmic models of belief revision, the latter serves as a logical description of the properties that any such operations should satisfy. The two can also be combined. An algorithmic contraction procedure may be checked for correctness according to given postulates. (Say, one which states that the result of contracting Θ with φ should be included in the original state $(\Theta - \varphi \subseteq \Theta.)$).

The last aspect (c) concerns the epistemic quality to be preserved. While the *foundationalists* argue that beliefs must be justified (with the exception of a selected set of 'basic beliefs'), the *coherentists* consider it a priority to maintain the overall coherence of the system and reject the existence of basic beliefs.

Therefore, each theory of epistemic change may be characterized by its representation of belief states, its description of belief revision operations, and its stand on the main properties of sets of beliefs one should be looking for. These choices may be interdependent. Say, a constructive approach might favor a representation by belief bases, and hence define belief revision operations on some finite base, rather than the whole background theory. Moreover, the epistemological stance determines what constitutes *rational epistemic change*. The foundationalist accepts only those beliefs that are justified in virtue of other basic beliefs, thus having an additional challenge of computing the reasons for an incoming belief. On the other hand, the coherentist must maintain coherence, and hence make only those minimal changes that do not endanger (at least) consistency (however, coherence need not be identified with consistency).

In particular, the AGM paradigm represents belief states as sets (in fact, as theories closed under logical consequence), provides 'rationality postulates' to characterize the belief revision operations (cf. postulates for contraction at the end of the chapter), and finally, it advocates a coherentist view. The latter is based on the empirical claim that people do not keep track of justifications for their beliefs, as some psychological experiments seem to indicate [Har65].

Abduction as Belief Revision

Abductive reasoning may be seen as an epistemic process for belief revision. In this context an incoming sentence φ is not necessarily an observation, but rather a belief for which an explanation is sought. Existing approaches to abduction usually do not deal with the issue of incorporating φ into the set of beliefs. Their concern is just how to give an account for φ. If the underlying theory is closed under logical consequence, however, then φ should be automatically added once we have added its explanation (which a foundationalist would then keep tagged as such).

In AI, practical connections of abduction to theories of belief revision have often been noted. But these in general use abduction to determine explanations of incoming beliefs or as an aid to perform the epistemic operations for belief revision. Of many references in the literature, we mention [Wil94] (which studies the relationship between explanations based on abduction and 'Spohnian reasons') and [AD94](which uses abductive procedures to realize contractions over theories with 'immutability conditions'). Some work has been done relating abduction to knowledge assimilation [KM90], in which the goal is to assimilate a series of observations into a theory maintaining its consistency.

Our claim will be stronger. Abduction can function in a model of theory revision as a means to determine explanations for incoming beliefs. But also more generally, abductive reasoning itself provides a model for epistemic change. Let us discuss some reasons for this, recalling our architecture of chapter 2.

First, what were called the two 'triggers' for abductive reasoning correspond to the two epistemic attitudes of a formula being undetermined or rejected. We did not consider accepted beliefs, since these do not call for explanation.

- φ is a novelty ($\Theta \not\models \varphi, \Theta \not\models \neg\varphi$): φ is undetermined.
- φ is an anomaly ($\Theta \not\models \varphi, \Theta \models \neg\varphi$: φ is rejected.
- φ is an accepted belief ($\Theta \models \varphi$).

The epistemic attitudes are presented in [Gar88] in terms of membership (e.g., a formula φ is accepted if $\varphi \in \Theta$). We defined them in terms of entailment, since our theories are not closed under logical consequence.

Our main concern is not the incoming belief φ itself. We rather want to compute and add its explanation α. But since φ is a logical consequence of the revised theory, it could easily be added. Thus, as we shall see, abduction as epistemic change may be described by two operations: either as an (abductive) expansion, where the background theory gets extended to account for a *novel* fact, or as an (abductive) revision, in which the theory needs to be revised to account for an *anomalous* fact. Belief revision theories provide an explicit calculus of modification for both cases and indeed serve as a guide to define abductive operations for epistemic change. In AI, characterizing abduction via belief operators goes back to Levesque [Lev89]. More recently, other work has been done in this direction ([Pag96], [LU96]). As we shall see, while the approaches share the same intuition, they all differ in implementation.

In philosophy, the idea of abduction as epistemic change is already present in Peirce's philosophical system (cf. chapter 7), in the connection between abduction and the epistemic transition between the mental states of doubt and belief. It shows itself very clearly in the fact that *surprise* is both the trigger of abductive reasoning, —as indicated by the first premise of the logical formulation–, as well as the trigger of the state of doubt, when a belief habit has been broken. Abductive reasoning is a process by which doubts are transformed into *beliefs*, since an explanation for a surprising fact is but a suggestion to be tested thereafter. Moreover, one aspect of abduction is related to the "Ramsey test" [Ram00]: given a conditional sentence $\alpha \rightarrow \varphi$, α is a reason for φ iff revising your current beliefs by α *causes* φ to be believed. This test has been the point of departure of much epistemological work both in philosophy and in AI.

Next we propose two abductive epistemic operations for the acquisition of knowledge, following those in the AGM theory while being faithful to our interpretation of Peirce's abduction and proposed taxonomy. Then, a discussion on abductive epistemic theories, in which we sketch ours and compare it to other work.

Abductive Operations for Epistemic Change

The previously defined abductive novelty and abductive anomaly correspond respectively, to the AGM epistemic attitudes of undetermination and rejection (provided that $\Rightarrow$ is $\vdash$ and Θ closed under logical consequence).

In our account of abduction, both a novel phenomenon and an anomalous one induce a change in the original theory. The latter calls for a revision and the former for expansion. So, the basic operations for abduction are expansion and revision. Therefore, two epistemic attitudes and changes in them are reflected in an abductive model.

Here, then, are the abductive operations for epistemic change:

- Abductive Expansion

  Given an abductive novelty φ, a consistent explanation α for φ is computed in such a way that $\Theta, \alpha \Rightarrow \varphi$, and then added to Θ.

- Abductive Revision

  Given an abductive anomaly φ, a consistent explanation α is computed as follows: the theory Θ is revised into Θ' so that it does not explain $\neg\varphi$. That is, $\Theta' \not\Rightarrow \neg\varphi$, where $\Theta' = \Theta - (\beta_1, \ldots, \beta_l)$[3].

  Once Θ' is obtained, a consistent explanation α is calculated in such a way that $\Theta', \alpha \Rightarrow \varphi$ and then added to Θ.

  Thus, the process of revision involves both contraction and expansion.

In one respect, these operations are more general than their counterparts in the AGM theory, since incoming beliefs are incorporated into the theory **together** with their explanation (when the theory is closed under logical consequence). But in the type of sentences to accept, they are more restrictive. Given that in our model non-surprising facts (where $\Theta \Rightarrow \varphi$) are not candidates for being explained, abductive expansion does not apply to already accepted beliefs, and similarly, revision only accepts rejected facts. Other approaches however, do not commit themselves to the preconditions of novelty and anomaly that we have set forward. Pagnucco's abductive expansion [Pag96] is defined for an inconsistent input, but in this case the resulting state stays the same. Lobo and Uzcátegui's abductive expansion [LU96] is even closer to standard AGM expansion; it is in fact the same when every atom is "abducible".

Abductive Epistemic Theories

Once we interpret abductive reasoning as a model for epistemic change, in lines of those proposed in AI, the next question is: what kind of theory is an abductive epistemic theory?

As we saw previously, there are several choices for representing belief states, for characterizing the operations for epistemic change and finally, an epistemic theory adheres to either the foundationalist or the coherentist trend. The predominant line of what we have examined so far, is to stay close to the AGM approach. That is, to represent belief states as sets (in fact as closed theories), and to characterize the abductive operations of extension and revision through definitions, rationality postulates and a number of constructions motivated by those for AGM contraction and revision. As for epistemological stance, Pagnucco is careful to keep his proposal away from being interpreted as foundationalist; he thinks that having a special set of beliefs like the "abducibles", as found in [LU96], is *"against the coherentist spirit of the AGM"* [Pag96, page 174].

[3]In many cases, several formulas and not just one must be removed from the theory. The reason is that sets of formulas which entail (explain) φ should be removed. E.g., given $\Theta = \{\alpha \rightarrow \beta, \alpha, \beta\}$ and $\varphi = \neg\beta$, in order to make $\Theta, \neg\beta$ consistent, one needs to remove either $\{\beta, \alpha\}$ or $\{\beta, \alpha \rightarrow \beta\}$.

As for epistemological stance, here is our position. The main motivation for an abductive epistemic theory is to incorporate an incoming belief together with its explanation, the belief (or set of) that justifies it. This fact places abduction close to the above foundationalist line, which requires that beliefs are justified in terms of other basic beliefs. Often, abductive beliefs are used by a (scientific) community, so the earlier claim that individuals do not keep track of the justifications of their beliefs [Gar88] does not apply. On the other hand, an important feature of abductive reasoning is maintaining consistency of the theory. Otherwise, explanations would be meaningless (especially if $\Rightarrow$ is interpreted as classical logical consequence. C.f. chapter 3). Therefore, abduction is committed to the coherentist approach as well. This is not a case of opportunism. Abduction rather demonstrates that the earlier philosophical stances are not incompatible. Indeed, [Haa93] argues for an intermediate stance of 'foundherentism'. Combinations of foundationalist and coherentist approaches are also found in the AI literature [Gal92].

In our view, an abductive theory for epistemic change, which aims to model Peirce's abduction, naturally calls for a procedural approach. It should produce explanations for surprising phenomena and thus transform a state of doubt into one of belief. The AGM postulates describe expansions, contractions and revisions as epistemic *products* rather than processes in their own right. Their concern is with the nature of epistemic states, not with their dynamics. This gives them a 'static' flavour, which may not always be appropriate.

Therefore, in the following, we aim at giving a constructive model in which abduction is an epistemic activity.

3. Semantic Tableaux Revisited: Toward An Abductive Model for Belief Revision

The combination of stances that we just described naturally calls for a procedural approach to abduction as an activity. But then, the same motivations that we gave in chapter 4 apply. Semantic tableaux provided an attractive constructive representation of theories, and abductive expansion operations that work over them. So, here is a further challenge for this framework. Can we extend our abductive tableau procedures to also deal with revision?

What we need for this purpose is an account of contraction on tableaux. Revision will then be forthcoming through combination with expansion, as has been mentioned before.

Revision in Tableaux

Our main idea is extremely straightforward. In semantic tableaux, contraction of a theory Θ, so as to give up some earlier consequences, translates into the *opening* of a closed branch of $\mathcal{T}(\Theta)$. Let us explain this in more detail for the case of revision. The latter process starts with Θ, φ for which $\mathcal{T}(\Theta \cup \varphi)$ is closed. In order to revise Θ, the first goal is to stop $\neg\varphi$ from being a consequence of Θ. This is done by opening a closed branch of $\mathcal{T}(\Theta)$ not closed by φ, thus transforming it into $\mathcal{T}(\Theta')$. This first step solves the problem of retracting inconsistencies. The next step is (much as in chapter 4) to find an explanatory formula α for φ by extending the modified Θ' as to make it entail φ. Therefore, revising a theory in the tableau format can be formulated as a combination of two equally natural moves, namely, *opening* and *closing* of branches:

Given Θ, φ for which $\mathcal{T}(\Theta \cup \varphi)$ is closed, α is an abductive explanation if

1. There is a set of formulas $\beta_1, \ldots, \beta_l$ ($\beta_i \in \Theta$) such that

$\mathcal{T}(\Theta \cup \varphi) - (\beta_1, \ldots, \beta_l)$ is open.

Moreover, let $\Theta_1 = \Theta - (\beta_1, \ldots, \beta_l)$. We also require that

2. $\mathcal{T}((\Theta_1 \cup \neg\varphi) \cup \alpha)$ is closed.

How to implement this technically? To open a tableau, it may be necessary to retract several formulas $\beta_1, \ldots, \beta_l$ and not just one (cf. previous footnote). The second item in this formulation is precisely the earlier process of abductive extension, which has been developed in chapter 4. Therefore, from now on we concentrate on the first point of the above process, namely, how to contract a theory in order to restore consistency.

Our discussion will be informal. Through a series of examples, we discover several key issues of implementing contraction in tableaux. We explore some complications of the framework itself, as well as several strategies for restoring consistency, and the effects of these in the production of explanations for anomalous observations.

Contraction in Tableaux

The general case of contraction that we shall need is this. We have a currently inconsistent theory, of which we want to retain some propositions, and from which we want to reject some others. In the above case, the observed anom alous phenomenon was to be retained, while the throwaways were not given in advance, and must be computed by some algorithm. We start by discussing the less constrained case of any inconsistent theory, seeing how it may be made consistent through contraction, using its semantic tableau as a guide.

As is well–known, a contraction operation is not uniquely defined, as there may be several options for removing formulas from a theory Θ so as to restore consistency. Suppose $\Theta = \{p \wedge q, \neg p\}$. We can remove either $p \wedge q$ or $\neg p$ – the choice of which depends, as we have noticed, on preferential criteria aiming at performing a 'minimal change' over Θ.

We start by noting that opening a branch may not suffice for restoring consistency. Consider the following example.

Example 1

Let $\Theta = \{p \wedge q, \neg p, \neg q\}$

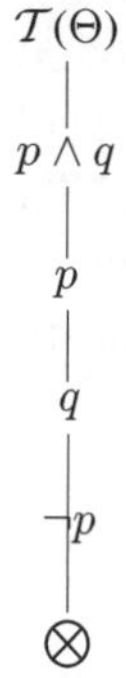

$$\begin{array}{c} \mathcal{T}(\Theta) \\ | \\ p \wedge q \\ | \\ p \\ | \\ q \\ | \\ \neg p \\ | \\ \otimes \end{array}$$

By removing $\neg p$ the closed branch is opened. However, note that this is not sufficient to restore consistency in Θ because $\neg q$ was never incorporated to the tableau! Thus, even upon removal of $\neg p$ from Θ, we have to 'recompute' the tableau, and we will find another closure, this time, because of $\neg q$. This phenomenon reflects a certain design decision for tableaux, which seemed harmless as long as we are merely testing for standard logical consequence. When constructing a tableau, as soon as a literal $\neg l$ may close a branch (i.e., l appears somewhere higher up; or vice versa) it does so, and no formula is added thereafter. Therefore, when opening a branch we are not sure that all formulas of the theory are represented on it. Thus, considerable reconfiguration (or even total reconstruction) may be needed before we can decide that a tableau has 'really' been opened. Of course (for our purposes), we might change the format of tableaux, and compute closed branches 'beyond inconsistency', so as to make all sources of closure explicit.

'Recomputation' is a complication arising from the specific tableau framework that we use, suggesting that we need to do more work in this setting than in other approaches to abduction. Moreover, it also illustrates that 'hidden conventions' concerning tableau construction may have unexpected effects, once we use tableaux for new purposes, beyond their original motivation. Granting all this, we feel that such phenomena are of some independent interest, and we continue with further examples demonstrating what tableaux have to offer for the study of contraction and restoring consistency.

Global and Local: Strategies for Contraction

Consider the following variant of our preceding example:

Example 2

Let $\Theta = \{p \wedge q, \neg p\}$

$$\begin{array}{c} \mathcal{T}(\Theta) \\ | \\ p \wedge q \\ | \\ p \\ | \\ q \\ | \\ \neg p \\ | \\ \otimes \end{array}$$

In this tableau, we can open the closed branch by removing either $\neg p$ or p. However, while $\neg p$ is indeed a formula of Θ, p is not. Here, if we follow standard accounts of contraction in theories of belief revision, we should trace back the Θ-source of this subformula ($p \wedge q$ in this case) and remove it. But tableaus offer another route. Alternatively, we could explore 'removing subformulas' from a theory by merely modifying their source formulas, as a more delicate kind of minimal change. These two alternatives suggest two strategies for contracting theories, which we label *global* and *local* contraction, respectively. Notice, in this connection, that each occurrence of a formula on a branch has a *unique history* leading up to one specific corresponding subformula occurrence in some formula from the theory Θ being analyzed by the tableau. The following illustrates what each strategy outputs as the contracted theory in our example:

- **Global Strategy**

  Branch–Opening $= \{\neg p, p\}$

  (i) Contract with $\neg p$:

  $\neg p$ corresponds to $\neg p$ in Θ

  $\Theta' = \Theta - \{\neg p\} = \{p \wedge q\}$

  (ii) Contract with p:

  p corresponds to $p \wedge q$ in Θ

  $\Theta' = \Theta - \{p \wedge q\} = \{\neg p\}$

- **Local Strategy**

  Branch–Opening $= \{\neg p, p\}$

(i) Contract with $\neg p$

Replace in the branch all connected occurrences of $\neg p$ (following its upward history) by the atom 'true': T.

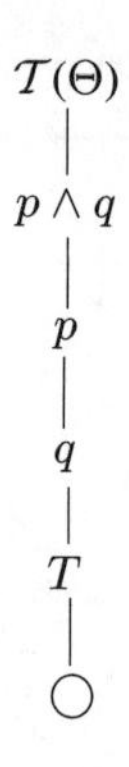

$\Theta' \;=\; \Theta - \{\neg p\} = \{p \wedge q, T\}$

(ii) Contract with p:

Replace in the branch all connected occurrences of p (following its history) by the atom true: T.

$\mathcal{T}(\Theta)$

$T \wedge q$

T

q

$\neg p$

$\bigcirc$

$\Theta' \;=\; \Theta - \{p\} = \{T \wedge q, \neg p\}$

Here, we have a case in which the two strategies differ. When contracting with p, the local strategy gives a revised theory (which is equivalent to $\{q, \neg p\}$) with less change than the global one. Indeed, if p is the source of inconsistency, why remove the whole formula $p \wedge q$ when we could modify it by $T \wedge q$? This simple example shows that 'removing subformulas' from branches, and modifying their source formulas, gives a more minimal change than removing the latter.

However, the choice is often less clear-cut. Sometimes the local strategy produces contracted theories that are logically equivalent to their globally contracted counterparts. Consider the following illustration (a variation on example 3 from chapter 2).

Example 3:

$\Theta = \{r \rightarrow l, r, \neg l\}$.

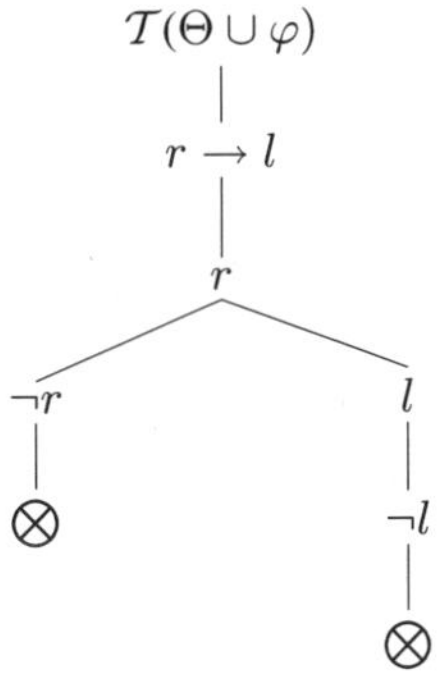

$\mathcal{T}(\Theta \cup \varphi)$

$r \rightarrow l$

r

$\neg r$ $\quad$ l

$\otimes$ $\quad$ $\neg l$

$\otimes$

Again, we briefly note the obvious outcomes of both local and global contraction strategies.

- **Global Strategy**

  Left Branch–Opening $= \{r, \neg r\}$

  (i) Contract with r:

  $\Theta' = \Theta - \{r\} = \{r \rightarrow l, \neg l\}$

  (ii) Contract with $\neg r$:

  $\Theta' = \Theta - \{r \rightarrow l\} = \{r, \neg l\}$

  Right Branch–Opening $= \{\neg l, l\}$

  (i) Contract with l:

  $\Theta' = \Theta - \{r \rightarrow l\} = \{r, \neg l\}$

  (ii) Contract with $\neg l$:

  $\Theta' = \Theta - \{\neg l\} = \{r, r \rightarrow l\}$

- **Local Strategy**

  Left Branch–Opening = $\{r, \neg r\}$

  (i) Contract with r:

  $\Theta' = \Theta - \{r\} = \{r \rightarrow l, \neg l\}$

  (ii) Contract with $\neg r$:

  $\Theta' = \{T \vee l, r, \neg l\}$.

  Right Branch–Opening = $\{r, \neg r\}$

  (i) Contract with l:

  $\Theta' = \{\neg r \vee T, r, \neg l\}$

  (ii) Contract with $\neg l$:

  $\Theta' = \Theta - \{\neg l\} = \{r, r \rightarrow l\}$

Now, the only deviant case in the local strategy is this. Locally contracting Θ with $\neg r$ makes the new theory $\{T \vee l, r, \neg l\}$. Given that the first formula is a tautology, the output is logically equivalent to its global counterpart $\{r, \neg l\}$. Therefore, modifying versus deleting conflicting formulae makes no difference in this whole example.

A similar indifference shows up in computations with simple disjunctions, although more complex theories with disjunctions of conjunctions may again show differences between the two strategies. We refrain from spelling out these examples here, which can easily be supplied by the reader. Also, we leave the exact domain of equivalence of the two strategies as an open question. Instead, we survey a useful practical case, again in the form of an example.

Computing Explanations

Let us now return to abductive explanation for an anomaly. We want to keep the latter fixed in what follows (it is precisely what needs to be accommodated), modifying merely the background theory. As it happens, this constraint involves just an easy modification of our contraction procedure so far.

Example 4

$\Theta = \{p \wedge q \rightarrow r, p \wedge q\}, \qquad \varphi = \neg r$

There are five possibilities on what to retract ($\neg r$ does not count since it is the anomalous observation). In the following descriptions of 'local output', note that a removed literal $\neg l$ will lead to a substitution of F (the 'falsum') for its source l in a formula of the input theory Θ.

- Contracting with p:

  Global Strategy: $\Theta' = \{p \wedge q \rightarrow r\}$

  Local Strategy: $\Theta' = \{p \wedge q \rightarrow r, T \wedge q\}$

- Contracting with $\neg p$:

  Global Strategy: Θ'=$\{p \wedge q\}$

  Local Strategy: Θ'=$\{F \wedge q \rightarrow r, p \wedge q\}$

- Contracting with q:

  Global Strategy: Θ'=$\{p \wedge q \rightarrow r\}$

  Local Strategy: Θ'=$\{p \wedge q \rightarrow r, p \wedge T\}$

- Contracting with $\neg q$:

  Global Strategy: Θ'=$\{p \wedge q\}$

  Local Strategy: Θ'=$\{p \wedge F \rightarrow r, p \wedge q\}$

- Contracting with r:

  Global Strategy: Θ'=$\{p \wedge q\}$

  Local Strategy: Θ'=$\{p \wedge q \rightarrow T, p \wedge q\}$.

A case in which the revised theories are equivalent is when contracting with r, so we have several cases in which we can compare the different explanations produced by our two strategies. To obtain the latter, we need to perform 'positive' standard abduction over the contracted theory. Let us look first at the case when the theory was contracted with p. Following the global strategy, the only explanation for $\varphi = \neg r$ with respect to the revised theory (Θ'=$\{p \wedge q \rightarrow r\}$) is the trivial solution, $\neg r$ itself. On the other hand, following the local strategy, there is another possible explanation for $\neg r$ with respect to its revised theory (Θ'=$\{p \wedge q \rightarrow r, T \wedge q\}$), namely $q \rightarrow \neg r$. Moreover, if we contract with $\neg p$, we get the same set of possible explanations in both strategies. Thus, again, the local strategy seems to allow for more 'pointed' explanations of anomalous observations.

We do not claim that either of our strategies is definitely better than the other one. We would rather point at the fact that tableaux admit of many plausible contraction operations, which we take to be a vindication of our framework. Indeed, tableaux also suggest a slightly more ambitious approach. We outline yet another strategy to restore consistency. It addresses a point mentioned in earlier chapters, viz. that explanation often involves changing one's 'conceptual framework'.

Contraction by Revising the Language

Suppose we have $\Theta = \{p \wedge q\}$, and we observe or learn that $\neg p$. Following our global contraction strategy would leave the theory empty, while following the local one would yield a contracted theory with $T \wedge q$ as its single formula. But there is another option, equally easy to implement in our tableau algorithms.

After all, in practice, we often resolve contradictions by 'making a distinction'. Mark the proposition inside the 'anomalous formula' ($\neg p$) by some *new proposition letter* (say p'), and replace its occurrences (if any) in the theory by the latter. In this case we obtain a new consistent theory Θ' consisting of $p \wedge q, \neg p$'. And other choice points in the above examples could be marked by suitable new proposition letters as well.

We may think of the pair p, p' as two variants of the same proposition, where some distinction has been made. Here is a simple illustration of this formal manipulation.

$p \wedge q$: "Rich and Famous"

$\neg p$: "Materially poor"

$\neg p$': "Poor in spirit"

In a dialogue, the 'anomalous statement' might then be defused as follows.

– A: "X is a rich and famous person, but X is poor."

– B: Why is X poor?"

Possible answers:

– A: "Because X is poor *in spirit*"

– A: "Because being rich makes X poor *in spirit*"

– A: "Because being famous makes X poor *in spirit*"

Over the new contracted (and reformulated) theories, our abductive algorithms of chapter 3 can easily produce these three consistent explanations (as $\neg p', p \rightarrow \neg p', q \rightarrow \neg p'$). The idea of reinterpreting the language to resolve inconsistencies suggests that there is more to belief revision and contraction than removing or modifying given formulas. The language itself may be a source of the anomaly, and hence *it* needs revision, too. (For related work in argumentation theory, cf. [vBe94].) This might be considered as a simple case of 'conceptual change'.

What we have shown is that, at least some language changes are easily incorporated into tableau algorithms, and are even suggested by them. Intuitively, any inconsistent theory can be made consistent by introducing enough distinctions into its vocabulary, and 'taking apart' relevant assertions. We must leave the precise extent, and algorithmic content, of this 'folklore fact' to further research.

Another appealing consequence of accommodating inconsistencies via language change, concerns structural rules of chapter 3. We will only mention that structural rules would acquire yet another parameter in their notation, namely the vocabulary over which the formulas are to be interpreted (e.g. $p|V_1, q|V_2 \Rightarrow p \wedge q|V_1 \cup V_2$). Interestingly, this format was also used in Bolzano [Bol73] (cf. chapter 3). An immediate side effect of this move are refined notions of consistency, in the spirit of those proposed by Hofstadter in [Hof79], in which consistency is relative to an 'interpretation'.

Outline of Contraction Algorithms

Global Strategy

Input: Θ, φ for which $\mathcal{T}(\Theta \cup \varphi)$ is closed.

Output: Θ' (Θ contracted) for which $\mathcal{T}(\Theta' \cup\varphi)$ is open.

Procedure: CONTRACT($\Theta, \neg\varphi, \Theta$')

Construct $\mathcal{T}(\Theta \cup \varphi)$, and label its closed branches: $\Gamma_1, \ldots, \Gamma_n$.

- IF $\neg\varphi \notin \Theta$
  Choose a closed branch Γ_i (not closed by $\varphi, \neg\varphi$).
  Calculate the literals that open it: Branch–Opening(Γ_i) $= \{\gamma_1, \gamma_2\}$. Choose one of them, say $\gamma = \gamma_1$.
  Find a corresponding formula γ' for γ in Θ higher up in the branch (γ' is either γ itself, or a formula in conjunctive or disjunctive form in which γ occurs.)
  Assign Θ' := $\Theta - \gamma$'.
- ELSE ($\neg\varphi \in \Theta$)
  Assign Θ' = $\Theta - \varphi$.
- IF $\mathcal{T}(\Theta'\cup\neg\varphi)$ is open AND all formulas from Θ are represented in the open branch, then go to END.
- ELSE
  IF $\mathcal{T}(\Theta'\cup\neg\varphi)$ is OPEN
  Add remaining formulas to the open branch(es) until there are no more formulas to add or until the tableau closes.
    IF the resulting tableau Θ" is open, reassign Θ':=Θ" and goto
    END.
    ELSE CONTRACT(Θ",$\neg\varphi$, Θ"').
  ELSE CONTRACT(Θ',$\neg\varphi$, Θ").
  % (This is the earlier-discussed 'iteration clause' for tableau recomputation.)
- END
  % (Θ' is the contracted theory with respect to $\neg\varphi$ such that $\mathcal{T}(\Theta' \cup\varphi)$ is open.)

Local Strategy

Input: Θ, φ for which $\mathcal{T}(\Theta \cup \varphi)$ is closed.

Output: Θ' (Θ contracted) for which $\mathcal{T}(\Theta' \cup\varphi)$ is open.

Procedure: CONTRACT($\Theta, \neg\varphi, \Theta$')

Construct $\mathcal{T}(\Theta \cup \varphi)$, and label its closed branches: $\Gamma_1, \ldots, \Gamma_n$.

- Choose a closed branch Γ_i (not closed by $\varphi, \neg\varphi$).
  Calculate the literals that open it: Branch–Opening(Γ_i) $= \{\gamma_1, \gamma_2\}$. Choose one of them, say $\gamma = \gamma_1$.
  Replace γ by T together with all its occurrences up in the branch. (γ' is either γ itself, or a formula in conjunctive or disjunctive form in which γ occurs.)
  Assign Θ' := $[T/\gamma]\Theta$.

- IF $\mathcal{T}$(Θ'∪¬φ) is open AND all formulas from Θ are represented in the open branch, then go to END.
- ELSE

  IF $\mathcal{T}$(Θ'∪¬φ) is OPEN

  Add remaining formulas to the open branch(es) until there are no more formulas to add or until the tableau closes.

    IF the resulting tableau Θ" is open, reassign Θ' := Θ" and goto

    END.

    ELSE CONTRACT(Θ",¬φ, Θ"').

  ELSE CONTRACT(Θ',¬φ, Θ").
- END

  % (Θ' is the contracted (by T substitution) theory with respect to ¬φ such that $\mathcal{T}$(Θ' ∪φ) is open.)

Rationality Postulates

To conclude our informal discussion of contraction in tableaux, we briefly discuss the AGM rationality postulates. (We list these postulates at the end of this chapter.) These are often taken to be the hallmark of any reasonable operation of contraction and revision – and many papers show laborious verifications to this effect. What do these postulates state in our case, and do they make sense?

To begin with, we recall that theories of epistemic change differed in the way their operations were defined (amongst other things). These can be given either 'constructively', as we have done, or via 'postulates'. The former procedures might then be checked for correctness according to the latter. However, in our case, this is not as straightforward as it may seem. The AGM postulates take belief states to be theories closed under logical consequence. But our tableaux analyze non-deductively closed finite sets of formulas, corresponding with 'belief bases'. This will lead to changes in the postulates themselves.

Here is an example. Postulate 3 for contraction says that: "If the formula to be retracted does not occur in the belief set K, nothing is to be retracted":

K-3 If $\varphi \notin K$, then $K - \varphi = K$.

In our framework, we cannot just replace belief states by belief bases here. Of course, the intuition behind the postulate is still correct. If φ is not a consequence of Θ (that we encounter in the tableau), then it will never be used for contraction by our algorithms. Another point of divergence is that our algorithms do not put the same emphasis on contracting one specific item from the background theory as the AGM postulates. This will vitiate further discussion of even more complex postulates, such as those breaking down contractions for complex formulas into successive cases.

One more general reason for this mismatch is the following. Despite their operational terminology (and ideology), the AGM postulates describe expansions, contractions, and revisions as (in the terminology of chapter 2) epistemic *products*, rather than *processes* in their own right. This gives them a 'static' flavor, which may not always be appropriate.

Therefore, we conclude that the AGM postulates as they stand do not seem to apply to contraction and revision procedures like ours. Evidently, this raises the issue of which general features *are* present in our algorithmic approach, justifying it as a legitimate notion of contraction. There is still a chance that a relatively slight 'revision of the revision postulates' will do the job. (An alternative more 'procedural' approach might be to view these issues rather in the *dynamic logic* setting of [dRi94], [vBe96a] .) We must leave this issue to further investigation.

4. Discussion and Conclusions

Belief Revision in Explanation

We present here an example which goes beyond our 'conservative algorithms' of chapter 4 and that illustrates abduction as a model for epistemic change. This example relates our medical diagnosis case given in chapter 2. As we have shown, computing explanations for incoming beliefs gives a richer model than many theories of belief revision, but this model is necessarily more complex. After all, it gives a much broader perspective on the processes of expansion and revision.

In standard belief revision in AI , given an undetermined belief (our case of novelty) the natural operation for modifying a theory is expansion. The reason is that the incoming belief is consistent with the theory, so the minimal change criterion dictates that it is enough to add it to the theory. Once abduction is considered however, the explanation for the fact has to be incorporated as well, and simple theory expansion might not be always appropriate. Consider our previous example of statistical reasoning in medical diagnosis (cf. chapter 2, and 5.2.4 of chapter 5), concerning the quick recovery of Jane Jones, which we briefly reproduce as follows:

$\Theta : L_1, L_2, L_3, C_1$[4]

$\varphi : E$

Given theory Θ, we want to explain why Jane Jones recovered quickly (φ). Clearly, the theory neither claims with high probability that she recov-

[4]Almost all cases of streptococcus infection clear up quickly after the administration of penicillin (L1). Almost no cases of penicillin-resistant streptococcus infection clear up quickly after the administration of penicillin (L2). Almost all cases of streptococcus infection clear up quickly after the administration of Belladonna, a homeopathic medicine (L3). Jane Jones had streptococcus infection (C1). Jane Jones recovered quickly (E).

ered quickly ($\Theta \not\Rightarrow \varphi$), nor that she did not ($\Theta \not\Rightarrow \neg\varphi$). We have a case of novelty, the observed fact is consistent with the theory. Now suppose a doctor comes with the following explanation for her quick recovery: "After careful examination, I have come to the conclusion that Jane Jones recovered quickly because although she received treatment with penicillin and was resistant, her grandmother had given her Belladonna". This is a perfectly sound and consistent explanation. However, note that having the fact that 'Jane Jones was resistant to penicillin' as part of the explanation does lower the probability of explaining her quick recovery, to the point of statistically implying the contrary. Therefore, in order to make sense of the doctor's explanation, the theory needs to be revised as well, deleting the statistical rule L_2 and replacing it with something along the following lines: "Almost no cases of penicillin-resistant streptococcus infection clear up quickly after the administration of penicillin, unless they are cured by something else" (L_2').

Thus, we have shown with this example that for the case of novelty in statistical explanation (which we reviewed in chapter 5), theory expansion may not be the appropriate operation to perform (let alone the minimal one), and theory revision might be the only way to salvage both consistency and high probability.

Is Abduction Belief Revision?

In the first part of this chapter we have argued for an epistemic model of abductive reasoning, in the lines of those proposed for belief revision in AI. However, this connection does not imply that abduction can be equated to belief revision. Let me discuss some reasons for this claim.

On the one hand, in its emphasis on explanations, an abductive model for epistemic change is richer than many theories of belief revision. Admittedly, though, not all cases of belief revision involve explanation, so the greater richness also reflects a restriction to a special setting. Moreover, in our model, not all input data is epistemically assimilated, but only that which is *surprising*; those facts which have not been explained by the theory, or that are in conflict with it. (Even so, one might speculate whether facts which are merely probable on the basis of Θ might still need explanation of some sort to further cement their status.)

On the other hand, having a model for abduction within the belief revision framework has imposed some other restrictions, which might not be appropriate in a broader conception of abductive reasoning. One is that our abduction always leads to some modification of the background theory to account for a surprising phenomenon. Therefore, we leave out cases in which a fact may be surprising (e.g. my computer is not working) even though it has been explained in the past. This sense of "surprising" seems closer to "unexpected"; there is a need to search for an explanation, but it does not involve any revision whatsoever.

Moreover, the type of belief revision accounted for in this model for abductive reasoning is of a very simple kind. Addition and removal of information are the basic operations to effect the epistemic changes, and the only criterion for theory revision is explanation. There is neither room for conceptual change, a more realistic type of scientific reasoning, nor place for a finer–grain distinction of kinds of expansion (as in [Lev91]) or of revision. Our approach may also be extended with a theory revision procedure, in which the revised theory is more *successful* than the original one, in the lines proposed in chapter 6 and in [Kui99].

Abductive Semantic Tableaux as a Framework for Epistemic Change

The second part of this chapter concerned our brief sketch of a possible use of semantic tableaux for performing all operations in an abductive theory of belief revision. Even in this rudimentary state, it presents some interesting features. For a start, expansion and contraction are not reverses of each other. The latter is essentially more complex. Expanding a tableau for Θ with formula φ merely hangs the latter to the open branches of $\mathcal{T}(\Theta)$. But retracting φ from Θ often requires complete reconfiguration of the initial tableau, and the contraction procedure needs to iterate, as a cascade of formulas may have to be removed. The advantage of contraction over expansion, however, is that we need not run any separate consistency checks, as we are merely weakening a theory.

We have not come down in favor of any of the strategies presented. The local strategy tends to retain more of the original theory, thus suggesting a more minimal change than the global one. Moreover, its 'substitution approach' is nicely in line with our alternative analysis of tableau abduction in section 4.5.2 of chapter 4, and it may lend itself to similar results. But in many practical cases, the more standard 'global abduction' works just as well. We must leave a precise comparison to future research.

Regarding different choices of formulas to be contracted, our algorithms are blatantly non-deterministic. If we want to be more focused, we would have to exploit the tableau structure itself to represent (say) some entrenchment order. The more fundamental formulas would lie closer to the root of the tree. In this way, instead of constructing all openings for each branch, we could construct only those closest to the leaves of the tree. (These correspond to the less important formulas, leaving the inner core of the theory intact.)

It would be of interest to have a proof-theoretic analysis of our contraction procedure. In a sense, the AGM 'rationality postulates' may be compared with the structural rules of chapter 3. And the more general question we have come up against is this: what are the appropriate logical constraints on a process view of contraction, and revision by abduction?

Conclusions

In this chapter we gave an account of abduction as a theory of belief revision. We proposed semantic tableaux as a logical representation of belief bases over which the major operations of epistemic change can be performed. The resulting theory combines the foundationalist with the coherentist stand in belief revision.

We claimed that beliefs need justification, and used our abductive machinery to construct explanations of incoming beliefs when needed. The result is richer than standard theories in AI, but it comes at the price of increased complexity. In fact, it has been claimed in [Doy92] that a realistic workable system for belief revision must not only trade deductive closed theories for belief bases, but also drop the consistency requirement. (As we saw in chapter 3, the latter is undecidable for sufficiently expressive predicate-logical languages. And it may still be NP–complete for sentences in propositional logic.) Given our analysis in chapter 3, we claim that any system, which aims at producing genuine explanations for incoming beliefs, must maintain consistency. What this means for workable systems of belief revision remains a moot point.

The tableau analysis confirms the intuition that revision is more complex than expansion, and that it admits of more variation. Several choices are involved, for which there seem to be various natural options, even in this constrained logical setting. What we have not explored in full is the way in which tableaux might generate entrenchment orders that we can profit from computationally. As things stand, different tableau procedures for revision may output very different explanations: abductive revision is not one unique procedure, but a family.

Even so, the preceding analysis may have shown that the standard logical tool of semantic tableaux has more uses than those for which they were originally designed.

Finally, regarding other epistemic operations in AI and their connection to abduction, we have briefly mentioned *update*, the process of keeping beliefs up–to–date as the world changes. Its connection to abduction points to an interesting area of research; the changing world might be full of new surprises, or existing beliefs might have lost their explanations. Thus, appropriate operations would have to be defined to keep the theory updated with respect to these changes.

Regarding the connection to other work, it would be interesting to compare our approach with that mentioned which follows the AGM line. Although this is not as straightforward as it may seem, we could at least check whether our algorithmic constructions of abductive expansion and revision validate the abductive postulates found in [Pag96] and [LU96].

AGM Postulates for Contraction

K-1 For any sentence ϕ and any belief set K, $K - \phi$ is a belief set.

K-2 No new beliefs occur in $K - \phi$: $K - \phi \subseteq K$.

K-3 If the formula to be retracted does not occur in the belief set, nothing is to be retracted:

If $\phi \notin K$, then $K - \phi = K$.

K-4 The formula to be retracted is not a logical consequence of the beliefs retained, unless it is a tautology:

If not $\vdash \phi$, then $\phi \notin K - \phi)$.

K-5 It is possible to undo contractions (Recovery Postulate):

If $\phi \in K$, then $K \subseteq (K - \phi) + \phi$.

K-6 The result of contracting logically equivalent formulas must be the same:

If $\vdash \phi \leftrightarrow \psi$, then $K - \phi = K - \psi$.

K-7 Two separate contractions may be performed by contracting the relevant conjunctive formula:

$K - \phi \cap K - \psi \subseteq K - \phi \wedge \psi$.

K-8 If $\phi \notin K - \phi \wedge \psi$, then $K - \phi \wedge \psi \subseteq K - \psi$.

References

[AAA90] *Automated Abduction. Working Notes.* 1990 Spring Symposium Series of the AAA. March 27–29. Stanford University. Stanford, CA. 1990.

[AGM85] C. Alchourrón, P. Gärdenfors, and D. Makinson. 'On the logic of theory change: Partial meet contraction and revision functions'. *Journal of Symbolic Logic*, 50: 510–530. 1985.

[AL96] N. Alechina and M. van Lambalgen. 'Generalized quantification as substructural logic'. *Journal of Symbolic Logic*, 61(3): 1006–1044. 1996.

[Ali96a] A. Aliseda. 'A unified framework for abductive and inductive reasoning in philosophy and AI', in *Abductive and Inductive Reasoning Workshop Notes*. pp. 1–6. European Conference on Artificial Intelligence (ECAI'96). Budapest. August, 1996.

[Ali96b] A. Aliseda. 'Toward a Logic of Abduction', in *Memorias del V Congreso Iberoamericano de Inteligencia Artifical (IBERAMIA'96)*, Puebla, México. October, 1996.

[Ali97] A. Aliseda. *Seeking Explanations: Abduction in Logic, Philosophy of Science and Artificial Intelligence*. PhD Dissertation, Philosophy Department, Stanford University. Published by the Institute for Logic, Language and Computation (ILLC), University of Amsterdam (ILLC Dissertation Series 1997–4). 1997.

[Ali98] A. Aliseda. 'Computing Abduction in Semantic Tableaux', *Computación y Sistemas: Revista Iberoamericana de Computación*, volume II, number 1, pp. 5–13. Centro de Investigación en Computación (CIC), Instituto Politécnico Nacional, México. D.F. 1998.

[Ali00a] A. Aliseda. 'Heurística, Hipótesis y Demostración en Matemáticas'. In A. Velasco (ed) *La Heurística: Un Concepto de las Ciencias y las Humanidades*, pp. 58–74. Colección Aprender a Aprender. Centro de Investigaciones Interdisciplinarias en Ciencias y Humanidades. UNAM, México. 2000.

[Ali00b] A. Aliseda. 'Abduction as Epistemic Change: A Peircean Model in Artificial Intelligence'. In P. Flach and A. Kakas (eds). *Abductive and Inductive Reason-*

ing: Essays on their Relation and Integration, pp. 45-58. Applied Logic Series. Kluwer Academic Publishers. 2000.

[Ali03a] A. Aliseda. 'Mathematical Reasoning Vs. Abductive Reasoning: An Structural Approach', *Synthese* 134: 25–44. Kluwer Academic Press. 2003.

[Ali03b] A. Aliseda. 'Abducción y Pragmati(ci)smo en C.S. Peirce'. In Cabanchik, S., etal. (eds) *El Giro Pragmático en la Filosofía Contemporánea*. Gedisa, Argentina. 2003.

[Ali04a] A. Aliseda. 'Logics in Scientific Discovery', *Foundations of Science*, Volume 9, Issue 3, pp. 339–363. Kluwer Academic Publishers. 2004.

[Ali04b] A. Aliseda. 'Sobre la Lógica del Descubrimiento Científico de Karl Popper', *Signos Filosóficos*, supplement to number 11, January–June, pp. 115–130. Universidad Autónoma Metropolitana. México. 2004.

[Ali05a] A. Aliseda. 'Lacunae, Empirical Progress and Semantic Tableaux'. In R. Festa, A. Aliseda and J. Peijnenburg (eds). *Confirmation, Empirical Progress, and Truth Approximation (Poznan Studies in the Philosophy of the Sciences and the Humanities*, vol. 83), pp. 141–161. Amsterdam/Atlanta, GA: Rodopi. 2005.

[Ali05b] A. Aliseda. 'The Logic of Abduction in The Light of Peirce's Pragmatism', *Semiotica*, 153–1/4, pp. 363–374. Semiotic Publications, Indiana University. 2005.

[AN04] A. Aliseda and A. Nepomuceno. *Abduction in First Order Semantic Tableaux*. Unpublished Manuscript. 2004.

[And86] D. Anderson. 'The Evolution of Peirce's Concept of Abduction'. *Transactions of the Charles S. Peirce Society 22-2:145–164*. 1986.

[And87] D. Anderson. *Creativity and the Philosophy of C.S. Peirce*. Martinus Nijhoff. Philosophy Library Volume 27. Martinus Nijhoff Publishers. 1987.

[Ant89] C. Antaki. 'Lay explanations of behaviour: how people represent and communicate their ordinary theories', in C. Ellis (ed), *Expert Knowledge and Explanation: The Knowledge–Language Interface*, pp. 201–211. Ellis Horwood Limited, England. 1989.

[AD94] C. Aravindan and P.M. Dung. 'Belief dynamics, abduction and databases'. In C. MacNish, D. Pearce, and L.M. Pereira (eds). *Logics in Artificial Intelligence. European Workshop JELIA'94*, pp. 66–85. Lecture Notes in Artificial Intelligence 838. Springer–Verlag. 1994.

[Ayi74] M. Ayim. 'Retroduction: The Rational Instinct'. *Transactions of the Charles S. Peirce Society 10-1:34–43*. 1974.

[BE93] J. Barwise and J. Etchemendy. *Tarski's World*. Center for the Study of Language and Information (CSLI). Stanford, CA. 1993.

[BE95] J. Barwise and J. Etchemendy. *Hyperproof*. Center for the Study of Language and Information (CSLI). Lecture Notes Series 42. Stanford, CA. 1995.

[Bat05] D. Batens. 'On a Logic of Induction'. In R. Festa, A. Aliseda and J. Peijnenburg (eds). *Confirmation, Empirical Progress, and Truth Approximation (Poznan Studies in the Philosophy of the Sciences and the Humanities*, vol. 83), pp. 195–221. Amsterdam/Atlanta, GA: Rodopi. 2005.

[vBe84a] J. van Benthem. 'Lessons from Bolzano'. Center for the Study of Language and Information. Technical Report CSLI-84-6. Stanford University. 1984. Later published as 'The variety of consequence, according to Bolzano'. *Studia Logica* 44, pp. 389–403. 1985.

[vBe84b] J. van Benthem. 'Foundations of Conditional Logic'. *Journal of Philosophical Logic'* 13, 303–349. 1984.

[vBe90] J. van Benthem. 'General Dynamics'. *ILLC Prepublication Series*, LP-90-11. Institute for Logic, Language, and Information. University of Amsterdam. 1990.

[vBe91] J. van Benthem. *Language in Action. Categories, Lambdas and Dynamic Logic*. Amsterdam: North Holland. 1991.

[vBe92] J. van Benthem. Logic as Programming. *Fundamenta Informaticae*, 17(4):285–318, 1992.

[vBe93] J. van Benthem. 'Logic and the Flow of Information'. *ILLC Prepublication Series* LP-91-10. Institute for Logic, Language, and Information. University of Amsterdam. 1993.

[vBe94] J. van Benthem. 'Logic and Argumentation'. *ILLC Research Report and Technical Notes Series* X-94-05. Institute for Logic, Language, and Information. University of Amsterdam. 1994.

[vBe96a] J. van Benthem. *Exploring Logical Dynamics*. CSLI Publications, Stanford University. 1996.

[vBe96b] J. van Benthem. Inference, Methodology, and Semantics. In P.I. Bystrov and V.N. Sadovsky (eds)., *Philosophical Logic and Logical Philosophy*, pp. 63–82. Kluwer Academic Publishers. 1996.

[vBC94] J. van Benthem, G. Cepparello. 'Tarskian Variations: Dynamic Parameters in Classical Semantics'. *Technical Report* CS–R9419. Centre for Mathematics and Computer Science (CWI), Amsterdam. 1994.

[Bet59] E.W. Beth. *The Foundations of Mathematics*. Amsterdam, North-Holland Publishing Company. 1959.

[Bet69] E. W. Beth. 'Semantic Entailment and Formal Derivability'. In J. Hintikka (ed), *The Philosophy of Mathematics*, pp. 9–41. Oxford University Press. 1969.

[Boo84] Boolos, G., 1984, 'Trees and finite satisfactibility', *Notre Dame Journal of Formal Logic* 25, pp. 110-115. 1984.

[BH02] I. C. Burger and J. Heidema. 'Degrees of Abductive Boldness'. Forthcoming. In L. Magnani, N. J. Nersessian, and C. Pizzi (eds)., *Logical and Computational Aspects of Model-Based Reasoning*, pp. 163–180. Kluwer Applied Logic Series. 2002.

[Bol73] B. Bolzano. *Wissenschaftslehre*, Seidel Buchhandlung, Sulzbach. Translated as *Theory of Science* by B. Torrel, edited by J. Berg. D. Reidel Publishing Company. Dordrecht, The Netherlands. 1973.

[BDK97] G. Brewka, J. Dix and K. Konolige. *Non-Monotonic Reasoning: An Overview*. Center for the study of Language and Information (CSLI), Lecture Notes 73. 1997.

[Bro68] B.A. Brody. 'Confirmation and Explanation', *Journal of Philosophy* 65, pp. 282-299. 1968.

[Car55] R. Carnap. *Statistical and Inductive Probability*. Galois Institute of Mathematics and Art. Brooklyn, New York. 1955.

[Cha97] G. Chaitin *The Limits of Mathematics: A Course on Information Theory and the Limits of Formal Reasoning*. Springer Verlag. 1997.

[Cho72] N. Chomsky. *Language and Mind. Enlarged Edition*. New York, Harcourt Brace Jovanovich. 1972.

[Cor75] J. Corcoran. 'Meanings of Implication', *Diaglos* 9, 59–76. 1975.

[DGHP99] D'Agostino, M., Gabbay, D.M., HÄhnle, R. and Posegga, J. (eds). *Handbook of Tableau Methods*. Kluwer Academic Publishers. 1999.

[Deb97] G. Debrock, "The artful riddle of abduction (abstract)", in M. Rayo etal. (eds.) *VIth International Congress of the International Association for Semiotic Studies: Semiotics Bridging Nature and Culture*. Editorial Solidaridad, México, p. 230. 1997.

[Dia93] E. Díaz. 'Arboles semánticos y modelos mínimos". In E. Pérez (ed.), *Actas del I Congreso de la Sociedad de Lógica, Metodología y Filosofía de la Ciencia en España*, pp. 40–43. Universidad Complutense de Madrid. 1993.

[Doy92] J. Doyle. 'Reason maintenance and belief revision: foundations versus coherence theories', in: P.Gärdenfors (ed), *Belief Revision*, pp. 29–51. Cambridge Tracts in Theoretical Computer Science, Cambridge University Press. 1992.

[DH93] K.Dosen and P. Schroeder-Heister (eds). *Substructural Logics*. Oxford Science Publications. Clarendon Press, Oxford. 1993.

[DP91b] D. Dubois and H. Prade. 'Possibilistic logic, preferential models, non-monotonicity and related issues', in *Proceedings of the 12th Inter. Joint Conf. on Artificial Intelligence (IJCAI–91)*, pp. 419–424. Sidney, Australia. 1991.

[ECAI96] *Abductive and Inductive Reasoning Workshop Notes*. European Conference on Artificial Intelligence (ECAI'96). Budapest. August, 1996.

[EG95] T. Eiter and G. Gottlob. 'The Complexity of Logic-Based Abduction', *Journal of the Association for Computing Machinery*, vol 42 (1): 3–42. 1995.

[Fan70] K.T.Fann. *Peirce's Theory of Abduction*. The Hague: Martinus Nijhoff. 1970.

[Fit90] M. Fitting. *First Order Logic and Automated Theorem Proving*. Graduate Texts in Computer Science. Springer–Verlag. 1990.

[Fla95] P.A. Flach. *Conjectures: an inquiry concerning the logic of induction*. PhD Thesis, Tilburg University. The Netherlands. 1995.

[Fla96a] P. Flach. 'Rationality Postulates for Induction', in Y. Shoham (ed), *Proceedings of the Sixth Conference on Theoretical Aspects of Rationality and Knowledge (TARK 1996)*, pp.267–281. De Zeeuwse Stromen, The Netherlands. March, 1996.

[Fla96b] P. Flach. 'Abduction and induction: syllogistic and inferential perspectives', in *Abductive and Inductive Reasoning Workshop Notes*. pp 31–35. European Conference on Artificial Intelligence (ECAI'96). Budapest. August, 1996.

[FK00] P. Flach and A. Kakas. *Abduction and Induction. Essays on their Relation and Integration*. Applied Logic Series. volume 18. Kluwer Academic Publishers. Dordrecht, The Netherlands. 2000.

[Fra58] H. Frankfurt. 'Peirce's Notion of Abduction'. *The Journal of Philosophy*, 55 (p.594). 1958.

[vFr80] B. van Fraassen. *The Scientific Image*. Oxford : Clarendon Press. 1980.

[Gab85] D.M. Gabbay. 'Theoretical foundations for non-monotonic reasoning in expert systems', in K. Apt (ed), *Logics and Models of Concurrent Systems*, pp. 439–457. Springer–Verlag. Berlin, 1985.

[Gab94a] D. M. Gabbay. 'Classical vs Non-classical Logics (The Universality of Classical Logic)', in D.M. Gabbay, C.J. Hogger and J.A. Robinson (eds), *Handbook of Logic in Artificial Intelligence and Logic Programming. Volume 2 Deduction Methodologies*, pp.359–500. Clarendon Press, Oxford Science Publications. 1994.

[Gab94b] D. M. Gabbay (ed). *What is a Logical System?*. Clarendon Press. Oxford. 1994.

[Gab96] D.M Gabbay. *Labelled Deductive Systems*. Oxford University Press. 1996.

[GJK94] D.M. Gabbay, R. Kempson. *Labeled Abduction and Relevance Reasoning*. Manuscript. Department of Computing, Imperial College and Department of Linguistics, School of Oriental and African Studies. London. 1994.

[GW05] D.M. Gabbay and J. Woods. *A Practical Logic of Cognitive Systems. The Reach of Abduction: Insight and Trial (volume 2)*. Amsterdam: North-Holland. 2005.

[Gal92] J.L. Galliers. 'Autonomous belief revision and communication', in: P.Gärdenfors (ed), *Belief Revision*, pp. 220–246. Cambridge Tracts in Theoretical Computer Science, Cambridge University Press, 1992.

[Gar88] P. Gärdenfors. *Knowledge in Flux: Modeling the Dynamics of Epistemic States*. MIT Press. 1988.

[Gar92] P. Gärdenfors. 'Belief revision: An introduction', in: P.Gärdenfors (ed), *Belief Revision*, pp. 1–28. Cambridge Tracts in Theoretical Computer Science, Cambridge University Press, 1992.

[GR95] P. Gärdenfors and H. Rott. 'Belief revision' in D.M. Gabbay, C.J. Hogger and J.A. Robinson (eds), *Handbook of Logic in Artificial Intelligence and Logic Programming*. Volume 4, Clarendon Press, Oxford Science Publications. 1995.

[GM88] P. Gärdenfors and David Makinson. 'Revisions of knowledge systems using epistemic entrenchment' in *Proceedings of the Second Conference on Theoretical Aspects of Reasoning and Knowledge*, pp 83–96. 1988.

[Ger95] P. Gervás. *Logical considerations in the interpretation of presuppositional sentences*. PhD Thesis, Department of Computing, Imperial College London. 1995.

[Gie91] R. Giere (ed). *Cognitive Models of Science*. Minnesota Studies in the Philosophy of Science, vol 15. Minneapolis: University of Minnesota Press. 1991.

[Gil96] D. Gillies. *Artificial Intelligence and Scientific Method*. Oxford University Press. 1996.

[Gin88] A. Ginsberg. 'Theory Revision via prior operationalization'. In *Proceedings of the Seventh Conference of the AAAI*. 1988.

[Gor97] D. Gorlée. '!Eureka! La traducción interlinguística como descubrimiento pragmático-abductivo (abstract)', in M. Rayo etal. (eds.) *VIth International Congress of the International Association for Semiotic Studies: Semiotics Bridging Nature and Culture*. Editorial Solidaridad, México, p. 230. 1997.

[Gro95] W. Groeneveld. *Logical Investigations into Dynamic Semantics*. PhD Dissertation, Institute for Logic, Language and Information, University of Amsterdam, 1995. (ILLC Dissertation Series 1995–18).

[Gut80] G. Gutting. 'The Logic of Invention', In T. Nickles (ed) *Scientific Discovery, Logic and Rationality*, 221–234. D. Reidel Publishing Company. 1980.

[Haa78] S. Haack. *Philosophy of Logics*. Cambridge University Press. New York. 1978. Blackwell, Oxford UK and Cambridge, Mass. 1993.

[Haa93] S. Haack. *Evidence and Inquiry. Towards Reconstruction in Epistemology*. Blackwell, Oxford UK and Cambridge, Mass. 1993.

[Han61] N.R. Hanson. *Patterns of Scientific Discovery*. Cambridge at The University Press. 1961.

[Har65] G. Harman. 'The Inference to the Best Explanation'. *Philosophical Review*. 74 88-95 (1965).

[Har86] G. Harman. *Change in View: Principles of Reasoning*. Cambridge, Mass. MIT Press, 1986.

[Hei83] I. Heim 'On the Projection Problem for Presuppositions', in *Proceedings of the Second West Coast Conference on Formal Linguistics, WCCFL*, volume 2: pp.114–26. 1983.

[Hel88] D. H. Helman (ed) *Analogical Reasoning : Perspectives of Artificial Intelligence, Cognitive Science and Philosophy*. Dordrecht, Netherlands: Reidel, 1988.

[Hem43] C. Hempel. 'A purely syntactical definition of confirmation', *Journal of Symbolic Logic* 6 (4):122–143. 1943.

[Hem45] C. Hempel. 'Studies in the logic of confirmation', *Mind* 54 (213): 1–26 (Part I); 54 (214): 97–121 (Part II). 1945.

[Hem65] C. Hempel. 'Aspects of Scientific Explanation', in C. Hempel *Aspects of Scientific Explanation and Other Essays in the Philosophy of Science*. The Free Press, New York. 1965.

[HO48] C. Hempel and P. Oppenheim. 'Studies in the logic of explanation', *Philosophy of Science* 15:135–175. 1948.

[Hin55] J. Hintikka. 'Two Papers on Symbolic Logic: Form and Quantification Theory and Reductions in the Theory of Types', *Acta Philosophica Fennica*, Fasc VIII, 1955.

[HR74] J. Hintikka and U. Remes. *The Method of Analysis: Its Geometrical Origin and Its General Significance*. D. Reidel Publishing Company. Dordrecht, Holland. 1974.

[HR76] J. Hintikka and U. Remes. 'Ancient Geometrical Analysis and Modern Logic'. In R.S. Cohen (ed), *Essays in Memory of Imre Lakatos*, pp. 253–276. D. Reidel Publishing Company. Dordrecht Holland. 1976.

[Hit95] C.R. Hitchcock. 'Discussion: Salmon on Explanatory Relevance', *Philosophy of Science*, pp. 304–320, 62 (1995).

[HSAM90] J.R. Hobbs, M. Stickel, D. Appelt, and P. Martin. 'Interpretation as Abduction'. *SRI International, Technical Note* 499. Artificial Intelligence Center, Computing and Engineering Sciences Division, Menlo Park, CA. 1990.

[Hof79] D.R. Hofstadter. *Gödel, Escher, Bach : An Eternal Golden Braid*. 1st Vintage Books ed. New York : Vintage Books. 1979.

[HHN86] J. Holland, K. Holyoak, R. Nisbett, R., and P. Thagard, *Induction: Processes of Inference, Learning, and Discovery*. Cambridge, MA:MIT Press/Bradford Books. 1986.

[Hoo92] C. Hookway. *Peirce*. Routledge, London 1992.

[Joh83] P.N. Johnson Laird. *Mental Models*. Cambridge, Mass. Harvard University Press. 1983.

[Jos94] J.R. Josephson. *Abductive Inference*. Cambridge University Press. 1994.

[KM90] A. Kakas and P. Mancarella. 'Knowledge Assimilation and Abduction'. In *Proceedings of the European Conference on Artificial Intelligence, ECAI'90*. International Workshop on Truth Maintenance. Springer–Verlag Lecture Notes in Computer Science. Stockholm. 1990.

[KKT95] A.C. Kakas, R.A.Kowalski, F. Toni. 'Abductive Logic Programming', *Journal of Logic and Computation* 2(6): pp. 719-770. 1995.

[Kal95] M. Kalsbeek. *Meta Logics for Logic Programming*. PhD Dissertation, Institute for Logic, Language and Information, University of Amsterdam, 1995. (ILLC Dissertation Series 1995–13).

[Kan93] M. Kanazawa. 'Completeness and Decidability of the Mixed Style of Inference with Composition'. Center for the Study of Language and Information (CSLI). *Technical Report* CSLI-93-181. Stanford University. Stanford, CA. 1993.

[Kap90] T. Kapitan. 'In What Way is Abductive Inference Creative?'. *Transactions of the Charles S. Peirce Society 26/4:499–512*. 1990.

[Ke89] K. Kelly. 'Effective Epistemology, Psychology, and Artificial Intelligence'. In W. Sieg (ed) *Acting and Reflecting. The Interdisciplinary Turn in Philosophy*, pp. 115–126. Synthese Library 211. Kluwer Academic Publishers. Dordrecht. 1989.

[Ke97] K. Kelly. *The Logic of Reliable Inquiry (Logic and Computation in Philosophy)*. Oxford University Press. 1997.

[Kon90] K. Konolige. 'A General Theory of Abduction'. In: *Automated Abduction, Working Notes*, pp. 62–66. Spring Symposium Series of the AAA. Stanford University. 1990.

[Kon96] K. Konolige. 'Abductive Theories in Artificial Intelligence'. In G. Brewka (ed), *Principles of Knowledge Representation*. Center for the Study of Language and Information (CSLI). Stanford University. Stanford, CA. 1996.

[Kow79] R.A. Kowalski. *Logic for problem solving*. Elsevier, New York, 1979.

[Kow91] R.A. Kowalski. 'A Metalogic Programming Approach to multiagent knowledge and belief', in V. Lifschitz (ed), *Artificial Intelligence and Mathematical Theory of Computation*, pp. 231–246. Academic Press, 1991.

[KLM90] S. Kraus, D. Lehmann, M. Magidor. 'Nonmonotonic Reasoning, Preferential Models and Cumulative Logics'. *Artificial Intelligence*, 44: 167–207. 1990.

[Kru95] G. Kruijff. *The Unbearable Demise of Surprise. Reflections on Abduction in Artificial Intelligence and Peirce's Philosophy*. Master Thesis. University of Twente. Enschede, The Netherlands. August, 1995.

[Kuh70] T. Kuhn. *The Structure of Scientific Revolutions* 2nd ed. Chicago: University of Chicago Press, 1970.

[Kui87] T.A.F. Kuipers (editor). *What is Closer-to-the-Truth?*, Rodopi, Amsterdam. 1987.

[Kui99] T. Kuipers. "Abduction Aiming at Empirical Progress or Even Truth Approximation Leading to a Challenge for Computational Modelling", in *Foundations of Science* 4: 307–323. 1999.

[Kui00] T. Kuipers. *From Instrumentalism to Constructive Realism. On some Relations between Confirmation, Empirical Progress and Truth Approximation*. Synthese Library 287, Kluwer AP, Dordrecht. The Netherlands. 2000.

[Kui01] T. Kuipers. *Structures in Science. Heuristic Patterns Based on Cognitive Structures*. Synthese Library 301, Kluwer AP, Dordrecht. The Netherlands. 2001.

[Kur95] N. Kurtonina. *Frames and Labels. A Modal Analysis of Categorial inference*. PhD Dissertation, Institute for Logic, Language and Information. University of Amsterdam, 1995. (ILLC Dissertation Series 1995–8).

[Lak76] I. Lakatos. *Proofs and Refutations. The Logic of Mathematical Discovery*. Cambridge University Press. 1976.

[Lau77] L. Laudan. *Progress and its Problems*. Berkeley, University of California Press. 1977.

[Lau80] L. Laudan. 'Why Was the Logic of Discovery Abandoned?', In T. Nickles (ed) *Scientific Discovery, Logic and Rationality*, pp. 173–183. D. Reidel Publishing Company. 1980.

[vLa96] M. van Lambalgen. 'Natural Deduction for Generalized Quantifiers'. In Jaap van der Does and Jan van Eijck (eds), *Quantifiers, Logic, and Language*, vol 54 of Lecture Notes, pp 225–236. Center for the Study of Language and Information (CSLI). Stanford, CA. 1996.

[Lap04] P.S. Laplace. 'Memoires'. In *Oeuvres complétes de Laplace*, vol. 13–14. Publiéees sous les auspices de l'Académie des sciences, par MM. les secrétaires perpétuels. Paris, Gauthier-Villars, 1878-1912.

[LSB87] P. Langley and H. Simon, G. Bradshaw, and J. Zytkow. *Scientific Discovery*. Cambridge, MA:MIT Press/Bradford Books. 1987.

[Lev89] H. J. Levesque. 'A knowledge-level accout of abduction', in *Proceedings of the eleventh International Joint Confererence on Artificial Intelligence*, pp. 1061–1067. 1989.

[Lev91] I. Levi. *The Fixation of Belief and its Undoing*. Cambridge University Press. 1991.

[LLo87] J.W. Lloyd. *Foundations of Logic Programming*. Springer–Verlag, Berlin, 2nd. edition. 1987.

[LU96] J. Lobo and C. Uzcategui. 'Abductive Change Operators', *Fundamenta Informaticae*, vol. 27, pp. 385–411. 1996.

[Mag01] L. Magnani. *Abduction, Reason, and Science: Processes of Discovery and Explanation*. Kluwer-Plenum, New York. 2001.

[Mak93] D. Makinson. 'General Patterns in Nonmonotonic Reasoning', in D.M. Gabbay, C.J. Hogger and J.A. Robinson (eds), *Handbook of Logic in Artificial Intelligence and Logic Programming. Volume 3, Nonmonoptonic reasoning and uncertain reasoning*, chapter 2.2. Clarendon Press, Oxford Science Publications. 1994.

[MP93] M.C. Mayer and F. Pirri. 'First order abduction via tableau and sequent calculi', in *Bulletin of the IGPL*, vol. 1, pp.99–117. 1993.

[MN99] J. Meheus and T. Nickles. 'The methodological study of discovery and creativity — Some background', *Foundations of Science* 4(3): pp. 231–235. 1999.

[Meh05] J. Meheus. 'Emprirical Progress and Ampliative Adaptive Logics'. In R. Festa, A. Aliseda and J. Peijnenburg (eds). *Confirmation, Empirical Progress, and Truth Approximation (Poznan Studies in the Philosophy of the Sciences and the Humanities*, vol. 83), pp. 167–191. Amsterdam/Atlanta, GA: Rodopi. 2005.

[Mey95] W.P. Meyer Viol *Instantial Logic. An Investigation into Reasoning with Instances*. Ph.D Dissertation, Institute for Logic, Language and Information, University of Amsterdam, 1995. (ILLC Dissertation Series 1995–11).

[McC80] J. McCarthy. 'Circumscription: A form of non-monotonic reasoning', *Artificial Intelligence* 13, pp. 27–39. 1980.

[Men64] E. Mendelson. *Introduction to Mathematical Logic*. New York, Van Nostrand. 1964.

[Mic94] R. Michalski. 'Inferential Theory of Learning: Developing Foundations for Multistrategy Learning' in *Machine Learning: A Multistrategy Approach*, Morgan Kaufman Publishers. 1994.

[Mill 58] J.S. Mill. *A System of Logic*. (New York, Harper & brothers, 1858). Reprinted in *The Collected Works of John Stuart Mill*, J.M Robson (ed), Routledge and Kegan Paul, London. 1958.

[Min90] G.E. Mints. 'Several formal systems of the logic programming', *Computers and Artificial Intelligence*, 9:19–41. 1990.

[Mus89] A. Musgrave. 'Deductive Heuristics', in K. Gavroglu etal. (eds). *Imre Lakatos and Theories of Scientific Change*, pp. 15–32. Kluwer Academic Publishers. 1989.

[Nag79] E. Nagel. *The Structure of Science. Problems in the Logic of Scientific Explanation*, 2nd edition. Hackett Publishing Company, Indianapolis, Cambridge. 1979.

[Nep99] A. Nepomuceno. 'Tablas semánticas y metalógica (El caso de la lógica de segundo orden)', *Crítica*, vol. XXXI, No. 93, pp. 21-47. 1999.

[Nep 02] A. Nepomuceno. 'Scientific Explanation and Modified Semantic Tableaux'. In L. Magnani, N. J. Nersessian, and C. Pizzi (eds) *Logical and Computational Aspects of Model-Based Reasoning*, pp. 181–198. Kluwer Applied Logic Series. Kluwer Academic Publishers. Dordrecht, The Netherlands. 2002.

[Nic80] T. Nickles. 'Introductory Essay: Scientific Discovery and the Future of Philosophy of Science'. In T. Nickles (ed) *Scientific Discovery, Logic and Rationality*, pp. 1–59. D. Reidel Publishing Company. 1980.

[Nie96] I. Niemelä. 'Implementing Circumpscription using a Tableau Method', in W. Wahlster (ed) *Proceedings of the 12th European Conference on Artificial Intelligence, ECAI'96*, pp 80–84. 1996.

[NT73] I. Niiniluoto and R. Tuomela. *Theoretical Concepts and Hypothetico-Deductive Inference*. D. Reidel Publishing Company. Dordrecht, Holland. 1973.

[Nii81] I. Niiniluoto. 'Statistical Explanation Reconsidered', *Synthese*, 48:437-72. 1981.

[Nii00] I. Niiniluoto. 'Hempel's Theory of Statistical Explanation'. In: J.H. Fetzer (ed)., *Science, Explanation, and Rationality: The Philosophy of Carl G. Hempel*. Oxford: Oxford University Press, pp. 138-163. 2000.

[Pag96] M. Pagnucco. *The Role of Abductive Reasoning within the Process of Belief Revision*. PhD Dissertation, Basser Department of Computer Science. University of Sidney. Australia. 1996.

[Pao02] F. Paoli. *Substructural logics : a primer*. Dordrecht : Kluwer Academic Publishers. Dordrecht, The Netherlands. 2002.

[Pau93] G. Paul. 'Approaches to Abductive Reasoning: An Overview'. *Artificial Intelligence Review*, 7:109–152, 1993.

[CP] C.S. Peirce. *Collected Papers of Charles Sanders Peirce*. Volumes 1–6 edited by C. Hartshorne, P. Weiss. Cambridge, Harvard University Press. 1931–1935; and volumes 7–8 edited by A.W. Burks. Cambridge, Harvard University Press. 1958.

[Pei55] C.S. Peirce. "The Fixation of Belief", in J. Buchler (ed) *Philosophical Writings of Peirce*. Dover Publications, Inc. 1955.

[PG87] D. Poole., R.G. Goebel., and Aleliunas. 'Theorist: a logical reasoning system for default and diagnosis', in Cercone and McCalla (eds), *The Knowledge Fronteer: Essays in the Representation of Knowledge*, pp. 331-352. Springer Verlag Lecture Notes in Computer Science 331-352. 1987.

[PR90] Y. Peng and J.A. Reggia. *Abductive Inference Models for Diagnostic Problem-Solving*. Springer Verlag. Springer Series in Symbolic Computation – Artificial Intelligence. 1990.

[PB83] J. Perry and J. Barwise. *Situations and Attitudes*. Cambridge, Mass: MIT Press. 1983.

[Pol45] G. Polya. *How to Solve it. A New Aspect of Mathematical Method*. Princeton University Press. 1945.

[Pol54] G. Polya. *Induction and Analogy in Mathematics*. Vol I. Princeton University Press. 1954.

[Pol62] G. Polya. *Mathematical Discovery. On Understanding, learning, and teaching problem solving*. Vol I. John Wiley & Sons, Inc. New York and London. 1962.

[Pop34] K. Popper. 'Scientific Method (1934)'. In D. Miller (ed) *A Pocket Popper*. Fontana Paperbacks. University Press, Oxford. 1983. This consists of the end of section 1, sections 2 and 3, and chapter II of [Pop59].

[Pop59] K. Popper. *The Logic of Scientific Discovery*. London: Hutchinson. (11th impression, 1979). 1959. (Originally published as *Logik der Forschung*, Springer. 1934).

[Pop60a] K. Popper. 'The Growth of Scientific Knowledge (1960)'. In D. Miller (ed) *A Pocket Popper*. Fontana Paperbacks. University Press, Oxford. 1983. This consists of Chapter 10 of [Pop63].

[Pop60b] K. Popper. 'Knowledge Without Authority (1960)'. In D. Miller (ed) *A Pocket Popper*. Fontana Paperbacks. University Press, Oxford. 1983. This consists of the introduction to [Pop63].

[Pop63] K. Popper. *Conjectures and Refutations. The Growth of Scientific Knowledge.* 5th ed. London and New York. Routledge. 1963.

[Pop75] K. Popper. *Objective Knowledge: An Evolutionary Approach*. Oxford University Press. 1975.

[Pop92] K. Popper. *Unended Quest: An Intellectual Autobiography*. Routledge, London. 1992.

[Pop73] H.E. Pople. 'On the Mechanization of Abductive Logic'. In: *Proceedings of the Third International Joint Conference on Artificial Intelligence* (IJCAI-73). San Mateo: Morgan Kauffmann, Stanford, CA. pp. 147–152. 1973.

[Qui61] W.V. Quine. 'On What there is', in *From a Logical Point of View*, 2nd ed.rev. Cambridge and Harvard University Press. 1961.

[Ram00] F. P. Ramsey. 'General Propositions and Causality', in R. B. Braithwaite (ed), *The Foundations of Mathematics and Other Logical Essays*. Routledge and Kegan Paul, pp. 237–257. 2000.

[Rei38] H. Reichenbach. *Experience and Prediction*. Chicago: University of Chicago Press. 1938.

[Rei70] F.E. Reilly. *Charles Peirce's Theory of Scientific Method*. New York: Fordham University Press, 1970.

[Rei80] R. Reiter. 'A Logic for default reasoning', *Artificial Intelligence* 13. 1980.

[Rei87] R. Reiter. 'A theory of diagnosis from first principles', *Artificial Intelligence* 32. 1987.

[Res78] N. Rescher. *Peirce's Philosophy of Science. Critical Studies in His Theory of Induction and Scientific Method.* University of Notre Dame. 1978.

[Rib96] P. Ribenboim. *The new book of prime number records*. 3rd edition. New York: Springer. 1996.

[dRi94] M. de Rijke. 'Meeting Some Neighbours', in J. van Eijck and A. Visser (eds) *Logic and Information Flow*. MIT Press, 1994.

[Roe97] A. Roesler. "Perception and Abduction (abstract)", in M. Rayo etal. (eds.) *VIth International Congress of the International Association for Semiotic Studies:*

Semiotics Bridging Nature and Culture. Editorial Solidaridad, México, p. 226. 1997.

[Rot88] R.J. Roth. 'Anderson on Peirce's Concept of Abduction'. *Transactions of the Charles S. Peirce Society 24/1:131–139*, 1988.

[Rya92] M. Ryan. *Ordered Presentation of Theories: Default Reasoning and Belief Revision*. PhD Thesis, Department of Computing, Imperial College London, 1992.

[Rub90] D-H. Ruben. *Explaining Explanation*. Routledge. London and New York. 1990.

[Sab90] M.R. Sabre. 'Peirce's Abductive Argument and the Enthymeme', *Transactions of the Charles S. Peirce Society*, 26/3:363–372. 1990.

[Sal71] W. Salmon. *Statistical Explanation and Statistical Relevance*. University of Pittsburgh Press, pp.29-87, 1971.

[Sal77] W. Salmon. 'A Third Dogma of Empiricism'. in Butts and Hintikka (eds)., *Basic Problems in Methodology and Linguistics*, Reidel, Dordrecht. 1977.

[Sal84] W. Salmon *Scientific Explanation and the Casual Structure of the World*. Princeton: Princeton University Press, 1984.

[Sal90] W. Salmon. *Four Decades of Scientific Explanation*. University of Minnesota Press, 1990.

[Sal92] W. Salmon. 'Scientific Explanation', in W. Salmon etal (eds)., *Introduction to the Philosophy of Science*. Prentice Hall, 1992.

[Sal94] W. Salmon. Causality without Counterfactuals, *Philosophy of Science* 61:297-312. 1994.

[Sav95] C. Savary. 'Discovey and its Logic: Popper and the "Friends of Discovery"'. *Philosophy of the Social Sciences*, Vol. 25., No. 3., 318–344. 1995.

[SD93] P. Schroeder-Heister and K. Dosen (eds.). *Substructural logics*. Oxford : Clarendon, 1993.

[Sco71] D. Scott. 'On Engendering an Illusion of Understanding, *Journal of Philosophy* 68, pp. 787–808. 1971.

[Sha91] E. Shapiro. 'Inductive Inference of Theories from Facts', in J. L. Lassez and G. Plotkin (eds), *Computational Logic: Essays in Honor of Alan Robinson.*. Cambridge, Mass. MIT, 1991.

[Sha70] R. Sharp. 'Induction, Abduction, and the Evolution of Science'. *Transactions of the Charles S. Peirce Society 6/1:17–33*. 1970.

[Sho88] Y. Shoham. *Reasoning about Change. Time and Causation from the standpoint of Artificial Intelligence*. The MIT Press, Cambridge, Mass, 1988.

[SL90] J. Shrager and P. Langley (eds). *Computational Models of Scientific Discovery and Theory Formation*. San Mateo: Morgan Kaufmann. 1990.

[SG73] H. Simon and G. Groen. 'Ramsey Eliminability and the Testability of Scientific Theories'. In H. Simon, *Models of Discovery*, pp. 403–421. A Pallas Paperback. Reidel, Holland. 1977. Originally published in *British Journal for the Philosophy of Science* 24, 357-408. 1973.

[Sim73a] H. Simon. 'Does Scientific Discovery Have a Logic?' In H. Simon, *Models of Discovery*, pp. 326–337. A Pallas Paperback. Reidel, Holland. 1977. (Originally published in *Philosophy of Science* 40, pp. 471–480. 1973.)

[Sim73b] H. Simon. 'The Structure of Ill-Structured Problems'. In H. Simon, *Models of Discovery*, pp. 304–325. A Pallas Paperback. Reidel, Holland. 1977. (Originally published in *Artificial Intelligence* 4, pp. 181–201. 1973).

[Sim77] H. Simon. *Models of Discovery*. A Pallas Paperback. Reidel, Holland. 1977.

[SLB81] H. Simon, P. Langley, and G. Bradshaw. 'Scientific Reasoning as Problem Solving', *Synthese* 47 pp. 1–27. 1981.

[SUM96] H. Sipma, T. Uribe, and Z. Manna. 'Deductive Model Checking'. Paper presented at the *Fifth CSLI Workshop on Logic, Language, and Computation*. Center for the Study of Language and Information (CSLI), Stanford University. June 1996.

[Smu68] R.M. Smullyan. *First Order Logic*. Springer Verlag, 1968.

[Sos75] E. Sosa, (ed). *Causation and Conditionals*. Oxford University Press. 1975.

[Ste83] W. Stegmüller. *Erklärung, Begründung, Kausalität*, second edition, Springer Verlag, Berlin, 1983.

[Sti91] M. Stickel. 'Upside-Down Meta-Interpretation of the Model Elimination Theorem-Proving Procedure for Deduction and Abduction'. *Technical Report* TR-664. Institute for New Generation Computer Technology (ICOT). 1991.

[Sup69] P.C. Suppes. *Studies in the methodology and foundations of science. selected papers from 1951 to 1969*. D. Reidel. Dordrecht. The Netherlans. 1969.

[Sup96] P. Suppes. *Foundations of Probability with Applications*. Cambridge University Press. 1996.

[Sza74] Á. K. Szabó *Working Backwards and Proving by Synthesis*. In [HR74, pp. 118–130]. 1974.

[Tam94] A.M. Tamminga, 'Logics of Rejection: Two Systems of Natural Deduction', in *Logique et analyse*, 37(146), p. 169. 1994.

[Tan92] Yao Hua Tan. *Non-Monotonic Reasoning: Logical Architecture and Philosophical Applications*. PhD Thesis, University of Amsterdam, 1992.

[Tar83] A. Tarski, *Logic, semantics, metamathematics : Papers from 1923 to 1938*. 2nd. edition. Indianapolis, Indiana : Hackett, 1983.

[Tha77] P.R. Thagard. The Unity of Peirce's Theory of Hypothesis. *Transactions of the Charles S. Peirce Society 13/2:112–123*. 1977.

[Tha88] P.R. Thagard. *Computational Philosophy of Science*. Cambridge, MIT Press. Bradford Books. 1988.

[Tha92] P.R. Thagard. *Conceptual Revolutions*. Princeton University Press. 1992.

[Tho81] P. Thompson. 'Bolzano's deducibility and Tarski's Logical Consequence', *History and Philosophy of Logic* 2, 11–20. 1981.

[Tij97] A. ten Teije *Automated Configuration of Problem Solving Methods in Diagnosis*. PhD Thesis, University of Amsterdam, 1997.

[Tou58] S. Toulmin. *The Uses of Argument*. Cambridge University Press, Cambridge, 1958.

[Tuo73] R.Tuomela. *Theoretical Concepts*. LEP Col. 10, Springer-Verlag, Vienna. 1973.

[Tro96] A.S. Troelstra. *Basic Proof Theory*. Cambridge University Press. 1996.

[VA99] A. Vázquez and A. Aliseda. 'Abduction in Semantic Tableaux: Towards An Object-Oriented Programming Implementation'. In J. M. Ahuactzin (ed.) *Encuentro Nacional de Computación, Taller de Lógica*. Sociedad Mexicana de Ciencia de la Computación. ISBN 968-6254-46-3. Hidalgo, México. September, 1999.

[Vis97] H. Visser. *Procedures of Discovery*. Manuscript. Katholieke Universiteit Brabant. Tilburg. The Netherlands. 1997.

[Wil94] M.A. Williams. 'Explanation and Theory Base Transmutations', in *Proceedings of the European Conference on Artificial Intelligence*, ECAI'94, pp. 341–246. 1994.

[Wir97] U. Wirth. 'Abductive inference in semiotics and philosophy of language: Peirce's and Davidson's account of interpretation (abstract)', in M. Rayo etal. (eds.) *VIth International Congress of the International Association for Semiotic Studies: Semiotics Bridging Nature and Culture*. Editorial Solidaridad, México, p. 232. 1997.

[Wri63] G.H. von Wright. *The Logic of Preference*. Edinburgh, University Press. 1963.

Author Index

Topic Index

SYNTHESE LIBRARY

1. J. M. Bochénski, *A Precis of Mathematical Logic.* Translated from French and German by O. Bird. 1959 ISBN 90-277-0073-7
2. P. Guiraud, *Problèmes et méthodes de la statistique linguistique.* 1959 ISBN 90-277-0025-7
3. H. Freudenthal (ed.), *The Concept and the Role of the Model in Mathematics and Natural and Social Sciences.* 1961 ISBN 90-277-0017-6
4. E. W. Beth, *Formal Methods.* An Introduction to Symbolic Logic and to the Study of Effective Operations in Arithmetic and Logic. 1962 ISBN 90-277-0069-9
5. B. H. Kazemier and D. Vuysje (eds.), *Logic and Language.* Studies dedicated to Professor Rudolf Carnap on the Occasion of His 70th Birthday. 1962 ISBN 90-277-0019-2
6. M. W. Wartofsky (ed.), *Proceedings of the Boston Colloquium for the Philosophy of Science, 1961–1962.* [Boston Studies in the Philosophy of Science, Vol. I] 1963 ISBN 90-277-0021-4
7. A. A. Zinov'ev, *Philosophical Problems of Many-valued Logic.* A revised edition, edited and translated (from Russian) by G. Küng and D.D. Comey. 1963 ISBN 90-277-0091-5
8. G. Gurvitch, *The Spectrum of Social Time.* Translated from French and edited by M. Korenbaum and P. Bosserman. 1964 ISBN 90-277-0006-0
9. P. Lorenzen, *Formal Logic.* Translated from German by F.J. Crosson. 1965 ISBN 90-277-0080-X
10. R. S. Cohen and M. W. Wartofsky (eds.), *Proceedings of the Boston Colloquium for the Philosophy of Science, 1962–1964.* In Honor of Philipp Frank. [Boston Studies in the Philosophy of Science, Vol. II] 1965 ISBN 90-277-9004-0
11. E. W. Beth, *Mathematical Thought.* An Introduction to the Philosophy of Mathematics. 1965 ISBN 90-277-0070-2
12. E. W. Beth and J. Piaget, *Mathematical Epistemology and Psychology.* Translated from French by W. Mays. 1966 ISBN 90-277-0071-0
13. G. Küng, *Ontology and the Logistic Analysis of Language.* An Enquiry into the Contemporary Views on Universals. Revised ed., translated from German. 1967 ISBN 90-277-0028-1
14. R. S. Cohen and M. W. Wartofsky (eds.), *Proceedings of the Boston Colloquium for the Philosophy of Sciences, 1964–1966.* In Memory of Norwood Russell Hanson. [Boston Studies in the Philosophy of Science, Vol. III] 1967 ISBN 90-277-0013-3
15. C. D. Broad, *Induction, Probability, and Causation.* Selected Papers. 1968 ISBN 90-277-0012-5
16. G. Patzig, *Aristotle's Theory of the Syllogism.* A Logical-philosophical Study of *Book A* of the *Prior Analytics.* Translated from German by J. Barnes. 1968 ISBN 90-277-0030-3
17. N. Rescher, *Topics in Philosophical Logic.* 1968 ISBN 90-277-0084-2
18. R. S. Cohen and M. W. Wartofsky (eds.), *Proceedings of the Boston Colloquium for the Philosophy of Science, 1966–1968, Part I.* [Boston Studies in the Philosophy of Science, Vol. IV] 1969 ISBN 90-277-0014-1
19. R. S. Cohen and M. W. Wartofsky (eds.), *Proceedings of the Boston Colloquium for the Philosophy of Science, 1966–1968, Part II.* [Boston Studies in the Philosophy of Science, Vol. V] 1969 ISBN 90-277-0015-X
20. J. W. Davis, D. J. Hockney and W. K. Wilson (eds.), *Philosophical Logic.* 1969 ISBN 90-277-0075-3
21. D. Davidson and J. Hintikka (eds.), *Words and Objections.* Essays on the Work of W. V. Quine. 1969, rev. ed. 1975 ISBN 90-277-0074-5; Pb 90-277-0602-6
22. P. Suppes, *Studies in the Methodology and Foundations of Science. Selected Papers from 1951 to 1969.* 1969 ISBN 90-277-0020-6
23. J. Hintikka, *Models for Modalities.* Selected Essays. 1969 ISBN 90-277-0078-8; Pb 90-277-0598-4

24. N. Rescher *et al.* (eds.), *Essays in Honor of Carl G. Hempel.* A Tribute on the Occasion of His 65th Birthday. 1969 ISBN 90-277-0085-0
25. P. V. Tavanec (ed.), *Problems of the Logic of Scientific Knowledge.* Translated from Russian. 1970 ISBN 90-277-0087-7
26. M. Swain (ed.), *Induction, Acceptance, and Rational Belief.* 1970 ISBN 90-277-0086-9
27. R. S. Cohen and R. J. Seeger (eds.), *Ernst Mach: Physicist and Philosopher.* [Boston Studies in the Philosophy of Science, Vol. VI]. 1970 ISBN 90-277-0016-8
28. J. Hintikka and P. Suppes, *Information and Inference.* 1970 ISBN 90-277-0155-5
29. K. Lambert, *Philosophical Problems in Logic.* Some Recent Developments. 1970 ISBN 90-277-0079-6
30. R. A. Eberle, *Nominalistic Systems.* 1970 ISBN 90-277-0161-X
31. P. Weingartner and G. Zecha (eds.), *Induction, Physics, and Ethics.* 1970 ISBN 90-277-0158-X
32. E. W. Beth, *Aspects of Modern Logic.* Translated from Dutch. 1970 ISBN 90-277-0173-3
33. R. Hilpinen (ed.), *Deontic Logic.* Introductory and Systematic Readings. 1971 *See also* No. 152. ISBN Pb (1981 rev.) 90-277-1302-2
34. J.-L. Krivine, *Introduction to Axiomatic Set Theory.* Translated from French. 1971 ISBN 90-277-0169-5; Pb 90-277-0411-2
35. J. D. Sneed, *The Logical Structure of Mathematical Physics.* 2nd rev. ed., 1979 ISBN 90-277-1056-2; Pb 90-277-1059-7
36. C. R. Kordig, *The Justification of Scientific Change.* 1971 ISBN 90-277-0181-4; Pb 90-277-0475-9
37. M. Čapek, *Bergson and Modern Physics.* A Reinterpretation and Re-evaluation. [Boston Studies in the Philosophy of Science, Vol. VII] 1971 ISBN 90-277-0186-5
38. N. R. Hanson, *What I Do Not Believe, and Other Essays.* Ed. by S. Toulmin and H. Woolf. 1971 ISBN 90-277-0191-1
39. R. C. Buck and R. S. Cohen (eds.), *PSA 1970.* Proceedings of the Second Biennial Meeting of the Philosophy of Science Association, Boston, Fall 1970. In Memory of Rudolf Carnap. [Boston Studies in the Philosophy of Science, Vol. VIII] 1971 ISBN 90-277-0187-3; Pb 90-277-0309-4
40. D. Davidson and G. Harman (eds.), *Semantics of Natural Language.* 1972 ISBN 90-277-0304-3; Pb 90-277-0310-8
41. Y. Bar-Hillel (ed.), *Pragmatics of Natural Languages.* 1971 ISBN 90-277-0194-6; Pb 90-277-0599-2
42. S. Stenlund, *Combinators, γ Terms and Proof Theory.* 1972 ISBN 90-277-0305-1
43. M. Strauss, *Modern Physics and Its Philosophy.* Selected Paper in the Logic, History, and Philosophy of Science. 1972 ISBN 90-277-0230-6
44. M. Bunge, *Method, Model and Matter.* 1973 ISBN 90-277-0252-7
45. M. Bunge, *Philosophy of Physics.* 1973 ISBN 90-277-0253-5
46. A. A. Zinov'ev, *Foundations of the Logical Theory of Scientific Knowledge (Complex Logic).* Revised and enlarged English edition with an appendix by G. A. Smirnov, E. A. Sidorenka, A. M. Fedina and L. A. Bobrova. [Boston Studies in the Philosophy of Science, Vol. IX] 1973 ISBN 90-277-0193-8; Pb 90-277-0324-8
47. L. Tondl, *Scientific Procedures.* A Contribution concerning the Methodological Problems of Scientific Concepts and Scientific Explanation. Translated from Czech by D. Short. Edited by R.S. Cohen and M.W. Wartofsky. [Boston Studies in the Philosophy of Science, Vol. X] 1973 ISBN 90-277-0147-4; Pb 90-277-0323-X
48. N. R. Hanson, *Constellations and Conjectures.* 1973 ISBN 90-277-0192-X

49. K. J. J. Hintikka, J. M. E. Moravcsik and P. Suppes (eds.), *Approaches to Natural Language.* 1973 ISBN 90-277-0220-9; Pb 90-277-0233-0
50. M. Bunge (ed.), *Exact Philosophy.* Problems, Tools and Goals. 1973 ISBN 90-277-0251-9
51. R. J. Bogdan and I. Niiniluoto (eds.), *Logic, Language and Probability.* 1973 ISBN 90-277-0312-4
52. G. Pearce and P. Maynard (eds.), *Conceptual Change.* 1973 ISBN 90-277-0287-X; Pb 90-277-0339-6
53. I. Niiniluoto and R. Tuomela, *Theoretical Concepts and Hypothetico-inductive Inference.* 1973 ISBN 90-277-0343-4
54. R. Fraissé, *Course of Mathematical Logic* – Volume 1: *Relation and Logical Formula.* Translated from French. 1973 ISBN 90-277-0268-3; Pb 90-277-0403-1
(For *Volume 2* see under No. 69).
55. A. Grünbaum, *Philosophical Problems of Space and Time.* Edited by R.S. Cohen and M.W. Wartofsky. 2nd enlarged ed. [Boston Studies in the Philosophy of Science, Vol. XII] 1973 ISBN 90-277-0357-4; Pb 90-277-0358-2
56. P. Suppes (ed.), *Space, Time and Geometry.* 1973 ISBN 90-277-0386-8; Pb 90-277-0442-2
57. H. Kelsen, *Essays in Legal and Moral Philosophy*. Selected and introduced by O. Weinberger. Translated from German by P. Heath. 1973 ISBN 90-277-0388-4
58. R. J. Seeger and R. S. Cohen (eds.), *Philosophical Foundations of Science.* [Boston Studies in the Philosophy of Science, Vol. XI] 1974 ISBN 90-277-0390-6; Pb 90-277-0376-0
59. R. S. Cohen and M. W. Wartofsky (eds.), *Logical and Epistemological Studies in Contemporary Physics.* [Boston Studies in the Philosophy of Science, Vol. XIII] 1973 ISBN 90-277-0391-4; Pb 90-277-0377-9
60. R. S. Cohen and M. W. Wartofsky (eds.), *Methodological and Historical Essays in the Natural and Social Sciences. Proceedings of the Boston Colloquium for the Philosophy of Science, 1969–1972.* [Boston Studies in the Philosophy of Science, Vol. XIV] 1974 ISBN 90-277-0392-2; Pb 90-277-0378-7
61. R. S. Cohen, J. J. Stachel and M. W. Wartofsky (eds.), *For Dirk Struik. Scientific, Historical and Political Essays*. [Boston Studies in the Philosophy of Science, Vol. XV] 1974 ISBN 90-277-0393-0; Pb 90-277-0379-5
62. K. Ajdukiewicz, *Pragmatic Logic*. Translated from Polish by O. Wojtasiewicz. 1974 ISBN 90-277-0326-4
63. S. Stenlund (ed.), *Logical Theory and Semantic Analysis.* Essays dedicated to Stig Kanger on His 50th Birthday. 1974 ISBN 90-277-0438-4
64. K. F. Schaffner and R. S. Cohen (eds.), *PSA 1972. Proceedings of the Third Biennial Meeting of the Philosophy of Science Association.* [Boston Studies in the Philosophy of Science, Vol. XX] 1974 ISBN 90-277-0408-2; Pb 90-277-0409-0
65. H. E. Kyburg, Jr., *The Logical Foundations of Statistical Inference.* 1974 ISBN 90-277-0330-2; Pb 90-277-0430-9
66. M. Grene, *The Understanding of Nature.* Essays in the Philosophy of Biology. [Boston Studies in the Philosophy of Science, Vol. XXIII] 1974 ISBN 90-277-0462-7; Pb 90-277-0463-5
67. J. M. Broekman, *Structuralism: Moscow, Prague, Paris*. Translated from German. 1974 ISBN 90-277-0478-3
68. N. Geschwind, *Selected Papers on Language and the Brain.* [Boston Studies in the Philosophy of Science, Vol. XVI] 1974 ISBN 90-277-0262-4; Pb 90-277-0263-2
69. R. Fraissé, *Course of Mathematical Logic* – Volume 2: *Model Theory.* Translated from French. 1974 ISBN 90-277-0269-1; Pb 90-277-0510-0
(For *Volume 1* see under No. 54)

SYNTHESE LIBRARY

70. A. Grzegorczyk, *An Outline of Mathematical Logic.* Fundamental Results and Notions explained with all Details. Translated from Polish. 1974 ISBN 90-277-0359-0; Pb 90-277-0447-3
71. F. von Kutschera, *Philosophy of Language.* 1975 ISBN 90-277-0591-7
72. J. Manninen and R. Tuomela (eds.), *Essays on Explanation and Understanding.* Studies in the Foundations of Humanities and Social Sciences. 1976 ISBN 90-277-0592-5
73. J. Hintikka (ed.), *Rudolf Carnap, Logical Empiricist.* Materials and Perspectives. 1975 ISBN 90-277-0583-6
74. M. Čapek (ed.), *The Concepts of Space and Time.* Their Structure and Their Development. [Boston Studies in the Philosophy of Science, Vol. XXII] 1976 ISBN 90-277-0355-8; Pb 90-277-0375-2
75. J. Hintikka and U. Remes, *The Method of Analysis.* Its Geometrical Origin and Its General Significance. [Boston Studies in the Philosophy of Science, Vol. XXV] 1974 ISBN 90-277-0532-1; Pb 90-277-0543-7
76. J. E. Murdoch and E. D. Sylla (eds.), *The Cultural Context of Medieval Learning.* [Boston Studies in the Philosophy of Science, Vol. XXVI] 1975 ISBN 90-277-0560-7; Pb 90-277-0587-9
77. S. Amsterdamski, *Between Experience and Metaphysics.* Philosophical Problems of the Evolution of Science. [Boston Studies in the Philosophy of Science, Vol. XXXV] 1975 ISBN 90-277-0568-2; Pb 90-277-0580-1
78. P. Suppes (ed.), *Logic and Probability in Quantum Mechanics.* 1976 ISBN 90-277-0570-4; Pb 90-277-1200-X
79. H. von Helmholtz: *Epistemological Writings. The Paul Hertz / Moritz Schlick Centenary Edition of 1921 with Notes and Commentary by the Editors.* Newly translated from German by M. F. Lowe. Edited, with an Introduction and Bibliography, by R. S. Cohen and Y. Elkana. [Boston Studies in the Philosophy of Science, Vol. XXXVII] 1975 ISBN 90-277-0290-X; Pb 90-277-0582-8
80. J. Agassi, *Science in Flux.* [Boston Studies in the Philosophy of Science, Vol. XXVIII] 1975 ISBN 90-277-0584-4; Pb 90-277-0612-2
81. S. G. Harding (ed.), *Can Theories Be Refuted?* Essays on the Duhem-Quine Thesis. 1976 ISBN 90-277-0629-8; Pb 90-277-0630-1
82. S. Nowak, *Methodology of Sociological Research.* General Problems. 1977 ISBN 90-277-0486-4
83. J. Piaget, J.-B. Grize, A. Szeminśska and V. Bang, *Epistemology and Psychology of Functions.* Translated from French. 1977 ISBN 90-277-0804-5
84. M. Grene and E. Mendelsohn (eds.), *Topics in the Philosophy of Biology.* [Boston Studies in the Philosophy of Science, Vol. XXVII] 1976 ISBN 90-277-0595-X; Pb 90-277-0596-8
85. E. Fischbein, *The Intuitive Sources of Probabilistic Thinking in Children.* 1975 ISBN 90-277-0626-3; Pb 90-277-1190-9
86. E. W. Adams, *The Logic of Conditionals.* An Application of Probability to Deductive Logic. 1975 ISBN 90-277-0631-X
87. M. Przełęcki and R. Wójcicki (eds.), *Twenty-Five Years of Logical Methodology in Poland.* Translated from Polish. 1976 ISBN 90-277-0601-8
88. J. Topolski, *The Methodology of History.* Translated from Polish by O. Wojtasiewicz. 1976 ISBN 90-277-0550-X
89. A. Kasher (ed.), *Language in Focus: Foundations, Methods and Systems.* Essays dedicated to Yehoshua Bar-Hillel. [Boston Studies in the Philosophy of Science, Vol. XLIII] 1976 ISBN 90-277-0644-1; Pb 90-277-0645-X

90. J. Hintikka, *The Intentions of Intentionality and Other New Models for Modalities.* 1975 ISBN 90-277-0633-6; Pb 90-277-0634-4
91. W. Stegmüller, *Collected Papers on Epistemology, Philosophy of Science and History of Philosophy.* 2 Volumes. 1977 Set ISBN 90-277-0767-7
92. D. M. Gabbay, *Investigations in Modal and Tense Logics with Applications to Problems in Philosophy and Linguistics.* 1976 ISBN 90-277-0656-5
93. R. J. Bogdan, *Local Induction.* 1976 ISBN 90-277-0649-2
94. S. Nowak, *Understanding and Prediction.* Essays in the Methodology of Social and Behavioral Theories. 1976 ISBN 90-277-0558-5; Pb 90-277-1199-2
95. P. Mittelstaedt, *Philosophical Problems of Modern Physics.* [Boston Studies in the Philosophy of Science, Vol. XVIII] 1976 ISBN 90-277-0285-3; Pb 90-277-0506-2
96. G. Holton and W. A. Blanpied (eds.), *Science and Its Public: The Changing Relationship.* [Boston Studies in the Philosophy of Science, Vol. XXXIII] 1976 ISBN 90-277-0657-3; Pb 90-277-0658-1
97. M. Brand and D. Walton (eds.), *Action Theory.* 1976 ISBN 90-277-0671-9
98. P. Gochet, *Outline of a Nominalist Theory of Propositions.* An Essay in the Theory of Meaning and in the Philosophy of Logic. 1980 ISBN 90-277-1031-7
99. R. S. Cohen, P. K. Feyerabend, and M. W. Wartofsky (eds.), *Essays in Memory of Imre Lakatos.* [Boston Studies in the Philosophy of Science, Vol. XXXIX] 1976 ISBN 90-277-0654-9; Pb 90-277-0655-7
100. R. S. Cohen and J. J. Stachel (eds.), *Selected Papers of Léon Rosenfield.* [Boston Studies in the Philosophy of Science, Vol. XXI] 1979 ISBN 90-277-0651-4; Pb 90-277-0652-2
101. R. S. Cohen, C. A. Hooker, A. C. Michalos and J. W. van Evra (eds.), *PSA 1974. Proceedings of the 1974 Biennial Meeting of the Philosophy of Science Association.* [Boston Studies in the Philosophy of Science, Vol. XXXII] 1976 ISBN 90-277-0647-6; Pb 90-277-0648-4
102. Y. Fried and J. Agassi, *Paranoia.* A Study in Diagnosis. [Boston Studies in the Philosophy of Science, Vol. L] 1976 ISBN 90-277-0704-9; Pb 90-277-0705-7
103. M. Przełęcki, K. Szaniawski and R. Wójcicki (eds.), *Formal Methods in the Methodology of Empirical Sciences.* 1976 ISBN 90-277-0698-0
104. J. M. Vickers, *Belief and Probability.* 1976 ISBN 90-277-0744-8
105. K. H. Wolff, *Surrender and Catch.* Experience and Inquiry Today. [Boston Studies in the Philosophy of Science, Vol. LI] 1976 ISBN 90-277-0758-8; Pb 90-277-0765-0
106. K. Kosík, *Dialectics of the Concrete.* A Study on Problems of Man and World. [Boston Studies in the Philosophy of Science, Vol. LII] 1976 ISBN 90-277-0761-8; Pb 90-277-0764-2
107. N. Goodman, *The Structure of Appearance.* 3rd ed. with an Introduction by G. Hellman. [Boston Studies in the Philosophy of Science, Vol. LIII] 1977 ISBN 90-277-0773-1; Pb 90-277-0774-X
108. K. Ajdukiewicz, *The Scientific World-Perspective and Other Essays, 1931-1963.* Translated from Polish. Edited and with an Introduction by J. Giedymin. 1978 ISBN 90-277-0527-5
109. R. L. Causey, *Unity of Science.* 1977 ISBN 90-277-0779-0
110. R. E. Grandy, *Advanced Logic for Applications.* 1977 ISBN 90-277-0781-2
111. R. P. McArthur, *Tense Logic.* 1976 ISBN 90-277-0697-2
112. L. Lindahl, *Position and Change.* A Study in Law and Logic. Translated from Swedish by P. Needham. 1977 ISBN 90-277-0787-1
113. R. Tuomela, *Dispositions.* 1978 ISBN 90-277-0810-X
114. H. A. Simon, *Models of Discovery and Other Topics in the Methods of Science.* [Boston Studies in the Philosophy of Science, Vol. LIV] 1977 ISBN 90-277-0812-6; Pb 90-277-0858-4

115. R. D. Rosenkrantz, *Inference, Method and Decision.* Towards a Bayesian Philosophy of Science. 1977 ISBN 90-277-0817-7; Pb 90-277-0818-5
116. R. Tuomela, *Human Action and Its Explanation.* A Study on the Philosophical Foundations of Psychology. 1977 ISBN 90-277-0824-X
117. M. Lazerowitz, *The Language of Philosophy*. Freud and Wittgenstein. [Boston Studies in the Philosophy of Science, Vol. LV] 1977 ISBN 90-277-0826-6; Pb 90-277-0862-2
118. Not published 119. J. Pelc (ed.), *Semiotics in Poland, 1894–1969.* Translated from Polish. 1979 ISBN 90-277-0811-8
120. I. Pörn, *Action Theory and Social Science.* Some Formal Models. 1977 ISBN 90-277-0846-0
121. J. Margolis, *Persons and Mind.* The Prospects of Nonreductive Materialism. [Boston Studies in the Philosophy of Science, Vol. LVII] 1977 ISBN 90-277-0854-1; Pb 90-277-0863-0
122. J. Hintikka, I. Niiniluoto, and E. Saarinen (eds.), *Essays on Mathematical and Philosophical Logic.* 1979 ISBN 90-277-0879-7
123. T. A. F. Kuipers, *Studies in Inductive Probability and Rational Expectation.* 1978 ISBN 90-277-0882-7
124. E. Saarinen, R. Hilpinen, I. Niiniluoto and M. P. Hintikka (eds.), *Essays in Honour of Jaakko Hintikka on the Occasion of His 50th Birthday.* 1979 ISBN 90-277-0916-5
125. G. Radnitzky and G. Andersson (eds.), *Progress and Rationality in Science.* [Boston Studies in the Philosophy of Science, Vol. LVIII] 1978 ISBN 90-277-0921-1; Pb 90-277-0922-X
126. P. Mittelstaedt, *Quantum Logic.* 1978 ISBN 90-277-0925-4
127. K. A. Bowen, *Model Theory for Modal Logic.* Kripke Models for Modal Predicate Calculi. 1979 ISBN 90-277-0929-7
128. H. A. Bursen, *Dismantling the Memory Machine.* A Philosophical Investigation of Machine Theories of Memory. 1978 ISBN 90-277-0933-5
129. M. W. Wartofsky, *Models*. Representation and the Scientific Understanding. [Boston Studies in the Philosophy of Science, Vol. XLVIII] 1979 ISBN 90-277-0736-7; Pb 90-277-0947-5
130. D. Ihde, *Technics and Praxis.* A Philosophy of Technology. [Boston Studies in the Philosophy of Science, Vol. XXIV] 1979 ISBN 90-277-0953-X; Pb 90-277-0954-8
131. J. J. Wiatr (ed.), *Polish Essays in the Methodology of the Social Sciences.* [Boston Studies in the Philosophy of Science, Vol. XXIX] 1979 ISBN 90-277-0723-5; Pb 90-277-0956-4
132. W. C. Salmon (ed.), *Hans Reichenbach: Logical Empiricist.* 1979 ISBN 90-277-0958-0
133. P. Bieri, R.-P. Horstmann and L. Krüger (eds.), *Transcendental Arguments in Science*. Essays in Epistemology. 1979 ISBN 90-277-0963-7; Pb 90-277-0964-5
134. M. Marković and G. Petrović (eds.), *Praxis*. Yugoslav Essays in the Philosophy and Methodology of the Social Sciences. [Boston Studies in the Philosophy of Science, Vol. XXXVI] 1979 ISBN 90-277-0727-8; Pb 90-277-0968-8
135. R. Wójcicki, *Topics in the Formal Methodology of Empirical Sciences.* Translated from Polish. 1979 ISBN 90-277-1004-X
136. G. Radnitzky and G. Andersson (eds.), *The Structure and Development of Science.* [Boston Studies in the Philosophy of Science, Vol. LIX] 1979 ISBN 90-277-0994-7; Pb 90-277-0995-5
137. J. C. Webb, *Mechanism, Mentalism and Metamathematics.* An Essay on Finitism. 1980 ISBN 90-277-1046-5
138. D. F. Gustafson and B. L. Tapscott (eds.), *Body, Mind and Method.* Essays in Honor of Virgil C. Aldrich. 1979 ISBN 90-277-1013-9
139. L. Nowak, *The Structure of Idealization.* Towards a Systematic Interpretation of the Marxian Idea of Science. 1980 ISBN 90-277-1014-7

140. C. Perelman, *The New Rhetoric and the Humanities.* Essays on Rhetoric and Its Applications. Translated from French and German. With an Introduction by H. Zyskind. 1979
ISBN 90-277-1018-X; Pb 90-277-1019-8
141. W. Rabinowicz, *Universalizability.* A Study in Morals and Metaphysics. 1979
ISBN 90-277-1020-2
142. C. Perelman, *Justice, Law and Argument.* Essays on Moral and Legal Reasoning. Translated from French and German. With an Introduction by H.J. Berman. 1980
ISBN 90-277-1089-9; Pb 90-277-1090-2
143. S. Kanger and S. Öhman (eds.), *Philosophy and Grammar.* Papers on the Occasion of the Quincentennial of Uppsala University. 1981 ISBN 90-277-1091-0
144. T. Pawlowski, *Concept Formation in the Humanities and the Social Sciences.* 1980
ISBN 90-277-1096-1
145. J. Hintikka, D. Gruender and E. Agazzi (eds.), *Theory Change, Ancient Axiomatics and Galileo's Methodology.* Proceedings of the 1978 Pisa Conference on the History and Philosophy of Science, Volume I. 1981 ISBN 90-277-1126-7
146. J. Hintikka, D. Gruender and E. Agazzi (eds.), *Probabilistic Thinking, Thermodynamics, and the Interaction of the History and Philosophy of Science.* Proceedings of the 1978 Pisa Conference on the History and Philosophy of Science, Volume II. 1981 ISBN 90-277-1127-5
147. U. Mönnich (ed.), *Aspects of Philosophical Logic.* Some Logical Forays into Central Notions of Linguistics and Philosophy. 1981 ISBN 90-277-1201-8
148. D. M. Gabbay, *Semantical Investigations in Heyting's Intuitionistic Logic.* 1981
ISBN 90-277-1202-6
149. E. Agazzi (ed.), *Modern Logic – A Survey.* Historical, Philosophical, and Mathematical Aspects of Modern Logic and Its Applications. 1981 ISBN 90-277-1137-2
150. A. F. Parker-Rhodes, *The Theory of Indistinguishables.* A Search for Explanatory Principles below the Level of Physics. 1981 ISBN 90-277-1214-X
151. J. C. Pitt, *Pictures, Images, and Conceptual Change.* An Analysis of Wilfrid Sellars' Philosophy of Science. 1981 ISBN 90-277-1276-X; Pb 90-277-1277-8
152. R. Hilpinen (ed.), *New Studies in Deontic Logic.* Norms, Actions, and the Foundations of Ethics. 1981 ISBN 90-277-1278-6; Pb 90-277-1346-4
153. C. Dilworth, *Scientific Progress.* A Study Concerning the Nature of the Relation between Successive Scientific Theories. 3rd rev. ed., 1994 ISBN 0-7923-2487-0; Pb 0-7923-2488-9
154. D. Woodruff Smith and R. McIntyre, *Husserl and Intentionality.* A Study of Mind, Meaning, and Language. 1982 ISBN 90-277-1392-8; Pb 90-277-1730-3
155. R. J. Nelson, *The Logic of Mind.* 2nd. ed., 1989 ISBN 90-277-2819-4; Pb 90-277-2822-4
156. J. F. A. K. van Benthem, *The Logic of Time.* A Model-Theoretic Investigation into the Varieties of Temporal Ontology, and Temporal Discourse. 1983; 2nd ed., 1991 ISBN 0-7923-1081-0
157. R. Swinburne (ed.), *Space, Time and Causality.* 1983 ISBN 90-277-1437-1
158. E. T. Jaynes, *Papers on Probability, Statistics and Statistical Physics.* Ed. by R. D. Rozenkrantz. 1983 ISBN 90-277-1448-7; Pb (1989) 0-7923-0213-3
159. T. Chapman, *Time: A Philosophical Analysis.* 1982 ISBN 90-277-1465-7
160. E. N. Zalta, *Abstract Objects.* An Introduction to Axiomatic Metaphysics. 1983
ISBN 90-277-1474-6
161. S. Harding and M. B. Hintikka (eds.), *Discovering Reality.* Feminist Perspectives on Epistemology, Metaphysics, Methodology, and Philosophy of Science. 1983
ISBN 90-277-1496-7; Pb 90-277-1538-6
162. M. A. Stewart (ed.), *Law, Morality and Rights.* 1983 ISBN 90-277-1519-X

163. D. Mayr and G. Süssmann (eds.), *Space, Time, and Mechanics.* Basic Structures of a Physical Theory. 1983 ISBN 90-277-1525-4
164. D. Gabbay and F. Guenthner (eds.), *Handbook of Philosophical Logic.* Vol. I: Elements of Classical Logic. 1983 ISBN 90-277-1542-4
165. D. Gabbay and F. Guenthner (eds.), *Handbook of Philosophical Logic.* Vol. II: Extensions of Classical Logic. 1984 ISBN 90-277-1604-8
166. D. Gabbay and F. Guenthner (eds.), *Handbook of Philosophical Logic.* Vol. III: Alternative to Classical Logic. 1986 ISBN 90-277-1605-6
167. D. Gabbay and F. Guenthner (eds.), *Handbook of Philosophical Logic.* Vol. IV: Topics in the Philosophy of Language. 1989 ISBN 90-277-1606-4
168. A. J. I. Jones, *Communication and Meaning.* An Essay in Applied Modal Logic. 1983 ISBN 90-277-1543-2
169. M. Fitting, *Proof Methods for Modal and Intuitionistic Logics.* 1983 ISBN 90-277-1573-4
170. J. Margolis, *Culture and Cultural Entities.* Toward a New Unity of Science. 1984 ISBN 90-277-1574-2
171. R. Tuomela, *A Theory of Social Action.* 1984 ISBN 90-277-1703-6
172. J. J. E. Gracia, E. Rabossi, E. Villanueva and M. Dascal (eds.), *Philosophical Analysis in Latin America.* 1984 ISBN 90-277-1749-4
173. P. Ziff, *Epistemic Analysis.* A Coherence Theory of Knowledge. 1984 ISBN 90-277-1751-7
174. P. Ziff, *Antiaesthetics.* An Appreciation of the Cow with the Subtile Nose. 1984 ISBN 90-277-1773-7
175. W. Balzer, D. A. Pearce, and H.-J. Schmidt (eds.), *Reduction in Science.* Structure, Examples, Philosophical Problems. 1984 ISBN 90-277-1811-3
176. A. Peczenik, L. Lindahl and B. van Roermund (eds.), *Theory of Legal Science.* Proceedings of the Conference on Legal Theory and Philosophy of Science (Lund, Sweden, December 1983). 1984 ISBN 90-277-1834-2
177. I. Niiniluoto, *Is Science Progressive?* 1984 ISBN 90-277-1835-0
178. B. K. Matilal and J. L. Shaw (eds.), *Analytical Philosophy in Comparative Perspective.* Exploratory Essays in Current Theories and Classical Indian Theories of Meaning and Reference. 1985 ISBN 90-277-1870-9
179. P. Kroes, *Time: Its Structure and Role in Physical Theories.* 1985 ISBN 90-277-1894-6
180. J. H. Fetzer, *Sociobiology and Epistemology.* 1985 ISBN 90-277-2005-3; Pb 90-277-2006-1
181. L. Haaparanta and J. Hintikka (eds.), *Frege Synthesized.* Essays on the Philosophical and Foundational Work of Gottlob Frege. 1986 ISBN 90-277-2126-2
182. M. Detlefsen, *Hilbert's Program.* An Essay on Mathematical Instrumentalism. 1986 ISBN 90-277-2151-3
183. J. L. Golden and J. J. Pilotta (eds.), *Practical Reasoning in Human Affairs.* Studies in Honor of Chaim Perelman. 1986 ISBN 90-277-2255-2
184. H. Zandvoort, *Models of Scientific Development and the Case of Nuclear Magnetic Resonance.* 1986 ISBN 90-277-2351-6
185. I. Niiniluoto, *Truthlikeness.* 1987 ISBN 90-277-2354-0
186. W. Balzer, C. U. Moulines and J. D. Sneed, *An Architectonic for Science.* The Structuralist Program. 1987 ISBN 90-277-2403-2
187. D. Pearce, *Roads to Commensurability.* 1987 ISBN 90-277-2414-8
188. L. M. Vaina (ed.), *Matters of Intelligence.* Conceptual Structures in Cognitive Neuroscience. 1987 ISBN 90-277-2460-1

189. H. Siegel, *Relativism Refuted.* A Critique of Contemporary Epistemological Relativism. 1987 ISBN 90-277-2469-5
190. W. Callebaut and R. Pinxten, *Evolutionary Epistemology.* A Multiparadigm Program, with a Complete Evolutionary Epistemology Bibliograph. 1987 ISBN 90-277-2582-9
191. J. Kmita, *Problems in Historical Epistemology.* 1988 ISBN 90-277-2199-8
192. J. H. Fetzer (ed.), *Probability and Causality.* Essays in Honor of Wesley C. Salmon, with an Annotated Bibliography. 1988 ISBN 90-277-2607-8; Pb 1-5560-8052-2
193. A. Donovan, L. Laudan and R. Laudan (eds.), *Scrutinizing Science.* Empirical Studies of Scientific Change. 1988 ISBN 90-277-2608-6
194. H.R. Otto and J.A. Tuedio (eds.), *Perspectives on Mind.* 1988 ISBN 90-277-2640-X
195. D. Batens and J.P. van Bendegem (eds.), *Theory and Experiment.* Recent Insights and New Perspectives on Their Relation. 1988 ISBN 90-277-2645-0
196. J. Österberg, *Self and Others.* A Study of Ethical Egoism. 1988 ISBN 90-277-2648-5
197. D.H. Helman (ed.), *Analogical Reasoning.* Perspectives of Artificial Intelligence, Cognitive Science, and Philosophy. 1988 ISBN 90-277-2711-2
198. J. Woleński, *Logic and Philosophy in the Lvov-Warsaw School.* 1989 ISBN 90-277-2749-X
199. R. Wójcicki, *Theory of Logical Calculi.* Basic Theory of Consequence Operations. 1988 ISBN 90-277-2785-6
200. J. Hintikka and M.B. Hintikka, *The Logic of Epistemology and the Epistemology of Logic.* Selected Essays. 1989 ISBN 0-7923-0040-8; Pb 0-7923-0041-6
201. E. Agazzi (ed.), *Probability in the Sciences.* 1988 ISBN 90-277-2808-9
202. M. Meyer (ed.), *From Metaphysics to Rhetoric.* 1989 ISBN 90-277-2814-3
203. R.L. Tieszen, *Mathematical Intuition.* Phenomenology and Mathematical Knowledge. 1989 ISBN 0-7923-0131-5
204. A. Melnick, *Space, Time, and Thought in Kant.* 1989 ISBN 0-7923-0135-8
205. D.W. Smith, *The Circle of Acquaintance.* Perception, Consciousness, and Empathy. 1989 ISBN 0-7923-0252-4
206. M.H. Salmon (ed.), *The Philosophy of Logical Mechanism.* Essays in Honor of Arthur W. Burks. With his Responses, and with a Bibliography of Burk's Work. 1990 ISBN 0-7923-0325-3
207. M. Kusch, *Language as Calculus vs. Language as Universal Medium.* A Study in Husserl, Heidegger, and Gadamer. 1989 ISBN 0-7923-0333-4
208. T.C. Meyering, *Historical Roots of Cognitive Science.* The Rise of a Cognitive Theory of Perception from Antiquity to the Nineteenth Century. 1989 ISBN 0-7923-0349-0
209. P. Kosso, *Observability and Observation in Physical Science.* 1989 ISBN 0-7923-0389-X
210. J. Kmita, *Essays on the Theory of Scientific Cognition.* 1990 ISBN 0-7923-0441-1
211. W. Sieg (ed.), *Acting and Reflecting.* The Interdisciplinary Turn in Philosophy. 1990 ISBN 0-7923-0512-4
212. J. Karpiński, *Causality in Sociological Research.* 1990 ISBN 0-7923-0546-9
213. H.A. Lewis (ed.), *Peter Geach: Philosophical Encounters.* 1991 ISBN 0-7923-0823-9
214. M. Ter Hark, *Beyond the Inner and the Outer.* Wittgenstein's Philosophy of Psychology. 1990 ISBN 0-7923-0850-6
215. M. Gosselin, *Nominalism and Contemporary Nominalism.* Ontological and Epistemological Implications of the Work of W.V.O. Quine and of N. Goodman. 1990 ISBN 0-7923-0904-9
216. J.H. Fetzer, D. Shatz and G. Schlesinger (eds.), *Definitions and Definability.* Philosophical Perspectives. 1991 ISBN 0-7923-1046-2
217. E. Agazzi and A. Cordero (eds.), *Philosophy and the Origin and Evolution of the Universe.* 1991 ISBN 0-7923-1322-4

SYNTHESE LIBRARY

218. M. Kusch, *Foucault's Strata and Fields.* An Investigation into Archaeological and Genealogical Science Studies. 1991 ISBN 0-7923-1462-X
219. C.J. Posy, *Kant's Philosophy of Mathematics.* Modern Essays. 1992 ISBN 0-7923-1495-6
220. G. Van de Vijver, *New Perspectives on Cybernetics.* Self-Organization, Autonomy and Connectionism. 1992 ISBN 0-7923-1519-7
221. J.C. Nyíri, *Tradition and Individuality.* Essays. 1992 ISBN 0-7923-1566-9
222. R. Howell, *Kant's Transcendental Deduction.* An Analysis of Main Themes in His Critical Philosophy. 1992 ISBN 0-7923-1571-5
223. A. García de la Sienra, *The Logical Foundations of the Marxian Theory of Value.* 1992 ISBN 0-7923-1778-5
224. D.S. Shwayder, *Statement and Referent.* An Inquiry into the Foundations of Our Conceptual Order. 1992 ISBN 0-7923-1803-X
225. M. Rosen, *Problems of the Hegelian Dialectic.* Dialectic Reconstructed as a Logic of Human Reality. 1993 ISBN 0-7923-2047-6
226. P. Suppes, *Models and Methods in the Philosophy of Science: Selected Essays.* 1993 ISBN 0-7923-2211-8
227. R. M. Dancy (ed.), *Kant and Critique: New Essays in Honor of W. H. Werkmeister.* 1993 ISBN 0-7923-2244-4
228. J. Woleński (ed.), *Philosophical Logic in Poland.* 1993 ISBN 0-7923-2293-2
229. M. De Rijke (ed.), *Diamonds and Defaults.* Studies in Pure and Applied Intensional Logic. 1993 ISBN 0-7923-2342-4
230. B.K. Matilal and A. Chakrabarti (eds.), *Knowing from Words.* Western and Indian Philosophical Analysis of Understanding and Testimony. 1994 ISBN 0-7923-2345-9
231. S.A. Kleiner, *The Logic of Discovery.* A Theory of the Rationality of Scientific Research. 1993 ISBN 0-7923-2371-8
232. R. Festa, *Optimum Inductive Methods.* A Study in Inductive Probability, Bayesian Statistics, and Verisimilitude. 1993 ISBN 0-7923-2460-9
233. P. Humphreys (ed.), *Patrick Suppes: Scientific Philosopher.* Vol. 1: Probability and Probabilistic Causality. 1994 ISBN 0-7923-2552-4
234. P. Humphreys (ed.), *Patrick Suppes: Scientific Philosopher.* Vol. 2: Philosophy of Physics, Theory Structure, and Measurement Theory. 1994 ISBN 0-7923-2553-2
235. P. Humphreys (ed.), *Patrick Suppes: Scientific Philosopher.* Vol. 3: Language, Logic, and Psychology. 1994 ISBN 0-7923-2862-0
     Set ISBN (Vols 233–235) 0-7923-2554-0
236. D. Prawitz and D. Westerståhl (eds.), *Logic and Philosophy of Science in Uppsala.* Papers from the 9th International Congress of Logic, Methodology, and Philosophy of Science. 1994 ISBN 0-7923-2702-0
237. L. Haaparanta (ed.), *Mind, Meaning and Mathematics.* Essays on the Philosophical Views of Husserl and Frege. 1994 ISBN 0-7923-2703-9
238. J. Hintikka (ed.), *Aspects of Metaphor.* 1994 ISBN 0-7923-2786-1
239. B. McGuinness and G. Oliveri (eds.), *The Philosophy of Michael Dummett.* With Replies from Michael Dummett. 1994 ISBN 0-7923-2804-3
240. D. Jamieson (ed.), *Language, Mind, and Art.* Essays in Appreciation and Analysis, In Honor of Paul Ziff. 1994 ISBN 0-7923-2810-8
241. G. Preyer, F. Siebelt and A. Ulfig (eds.), *Language, Mind and Epistemology.* On Donald Davidson's Philosophy. 1994 ISBN 0-7923-2811-6
242. P. Ehrlich (ed.), *Real Numbers, Generalizations of the Reals, and Theories of Continua.* 1994 ISBN 0-7923-2689-X

243. G. Debrock and M. Hulswit (eds.), *Living Doubt.* Essays concerning the epistemology of Charles Sanders Peirce. 1994 ISBN 0-7923-2898-1
244. J. Srzednicki, *To Know or Not to Know.* Beyond Realism and Anti-Realism. 1994 ISBN 0-7923-2909-0
245. R. Egidi (ed.), *Wittgenstein: Mind and Language.* 1995 ISBN 0-7923-3171-0
246. A. Hyslop, *Other Minds.* 1995 ISBN 0-7923-3245-8
247. L. Pólos and M. Masuch (eds.), *Applied Logic: How, What and Why.* Logical Approaches to Natural Language. 1995 ISBN 0-7923-3432-9
248. M. Krynicki, M. Mostowski and L.M. Szczerba (eds.), *Quantifiers: Logics, Models and Computation.* Volume One: Surveys. 1995 ISBN 0-7923-3448-5
249. M. Krynicki, M. Mostowski and L.M. Szczerba (eds.), *Quantifiers: Logics, Models and Computation.* Volume Two: Contributions. 1995 ISBN 0-7923-3449-3
Set ISBN (Vols 248 + 249) 0-7923-3450-7
250. R.A. Watson, *Representational Ideas from Plato to Patricia Churchland.* 1995 ISBN 0-7923-3453-1
251. J. Hintikka (ed.), *From Dedekind to Gödel.* Essays on the Development of the Foundations of Mathematics. 1995 ISBN 0-7923-3484-1
252. A. Wiśniewski, *The Posing of Questions.* Logical Foundations of Erotetic Inferences. 1995 ISBN 0-7923-3637-2
253. J. Peregrin, *Doing Worlds with Words.* Formal Semantics without Formal Metaphysics. 1995 ISBN 0-7923-3742-5
254. I.A. Kieseppä, *Truthlikeness for Multidimensional, Quantitative Cognitive Problems.* 1996 ISBN 0-7923-4005-1
255. P. Hugly and C. Sayward: *Intensionality and Truth.* An Essay on the Philosophy of A.N. Prior. 1996 ISBN 0-7923-4119-8
256. L. Hankinson Nelson and J. Nelson (eds.): *Feminism, Science, and the Philosophy of Science.* 1997 ISBN 0-7923-4162-7
257. P.I. Bystrov and V.N. Sadovsky (eds.): *Philosophical Logic and Logical Philosophy.* Essays in Honour of Vladimir A. Smirnov. 1996 ISBN 0-7923-4270-4
258. Å.E. Andersson and N-E. Sahlin (eds.): *The Complexity of Creativity.* 1996 ISBN 0-7923-4346-8
259. M.L. Dalla Chiara, K. Doets, D. Mundici and J. van Benthem (eds.): *Logic and Scientific Methods.* Volume One of the Tenth International Congress of Logic, Methodology and Philosophy of Science, Florence, August 1995. 1997 ISBN 0-7923-4383-2
260. M.L. Dalla Chiara, K. Doets, D. Mundici and J. van Benthem (eds.): *Structures and Norms in Science.* Volume Two of the Tenth International Congress of Logic, Methodology and Philosophy of Science, Florence, August 1995. 1997 ISBN 0-7923-4384-0
Set ISBN (Vols 259 + 260) 0-7923-4385-9
261. A. Chakrabarti: *Denying Existence.* The Logic, Epistemology and Pragmatics of Negative Existentials and Fictional Discourse. 1997 ISBN 0-7923-4388-3
262. A. Biletzki: *Talking Wolves.* Thomas Hobbes on the Language of Politics and the Politics of Language. 1997 ISBN 0-7923-4425-1
263. D. Nute (ed.): *Defeasible Deontic Logic.* 1997 ISBN 0-7923-4630-0
264. U. Meixner: *Axiomatic Formal Ontology.* 1997 ISBN 0-7923-4747-X
265. I. Brinck: *The Indexical 'I'.* The First Person in Thought and Language. 1997 ISBN 0-7923-4741-2
266. G. Hölmström-Hintikka and R. Tuomela (eds.): *Contemporary Action Theory.* Volume 1: Individual Action. 1997 ISBN 0-7923-4753-6; Set: 0-7923-4754-4

267. G. Hölmström-Hintikka and R. Tuomela (eds.): *Contemporary Action Theory.* Volume 2: Social Action. 1997 ISBN 0-7923-4752-8; Set: 0-7923-4754-4
268. B.-C. Park: *Phenomenological Aspects of Wittgenstein's Philosophy.* 1998 ISBN 0-7923-4813-3
269. J. Paśniczek: *The Logic of Intentional Objects.* A Meinongian Version of Classical Logic. 1998 Hb ISBN 0-7923-4880-X; Pb ISBN 0-7923-5578-4
270. P.W. Humphreys and J.H. Fetzer (eds.): *The New Theory of Reference.* Kripke, Marcus, and Its Origins. 1998 ISBN 0-7923-4898-2
271. K. Szaniawski, A. Chmielewski and J. Woleński (eds.): *On Science, Inference, Information and Decision Making.* Selected Essays in the Philosophy of Science. 1998 ISBN 0-7923-4922-9
272. G.H. von Wright: *In the Shadow of Descartes.* Essays in the Philosophy of Mind. 1998 ISBN 0-7923-4992-X
273. K. Kijania-Placek and J. Woleński (eds.): *The Lvov–Warsaw School and Contemporary Philosophy.* 1998 ISBN 0-7923-5105-3
274. D. Dedrick: *Naming the Rainbow.* Colour Language, Colour Science, and Culture. 1998 ISBN 0-7923-5239-4
275. L. Albertazzi (ed.): *Shapes of Forms.* From Gestalt Psychology and Phenomenology to Ontology and Mathematics. 1999 ISBN 0-7923-5246-7
276. P. Fletcher: *Truth, Proof and Infinity.* A Theory of Constructions and Constructive Reasoning. 1998 ISBN 0-7923-5262-9
277. M. Fitting and R.L. Mendelsohn (eds.): *First-Order Modal Logic.* 1998 Hb ISBN 0-7923-5334-X; Pb ISBN 0-7923-5335-8
278. J.N. Mohanty: *Logic, Truth and the Modalities from a Phenomenological Perspective.* 1999 ISBN 0-7923-5550-4
279. T. Placek: *Mathematical Intiutionism and Intersubjectivity.* A Critical Exposition of Arguments for Intuitionism. 1999 ISBN 0-7923-5630-6
280. A. Cantini, E. Casari and P. Minari (eds.): *Logic and Foundations of Mathematics.* 1999 ISBN 0-7923-5659-4 set ISBN 0-7923-5867-8
281. M.L. Dalla Chiara, R. Giuntini and F. Laudisa (eds.): *Language, Quantum, Music.* 1999 ISBN 0-7923-5727-2; set ISBN 0-7923-5867-8
282. R. Egidi (ed.): *In Search of a New Humanism.* The Philosophy of Georg Hendrik von Wright. 1999 ISBN 0-7923-5810-4
283. F. Vollmer: *Agent Causality.* 1999 ISBN 0-7923-5848-1
284. J. Peregrin (ed.): *Truth and Its Nature (if Any).* 1999 ISBN 0-7923-5865-1
285. M. De Caro (ed.): *Interpretations and Causes.* New Perspectives on Donald Davidson's Philosophy. 1999 ISBN 0-7923-5869-4
286. R. Murawski: *Recursive Functions and Metamathematics.* Problems of Completeness and Decidability, Gödel's Theorems. 1999 ISBN 0-7923-5904-6
287. T.A.F. Kuipers: *From Instrumentalism to Constructive Realism.* On Some Relations between Confirmation, Empirical Progress, and Truth Approximation. 2000 ISBN 0-7923-6086-9
288. G. Holmström-Hintikka (ed.): *Medieval Philosophy and Modern Times.* 2000 ISBN 0-7923-6102-4
289. E. Grosholz and H. Breger (eds.): *The Growth of Mathematical Knowledge.* 2000 ISBN 0-7923-6151-2

290. G. Sommaruga: *History and Philosophy of Constructive Type Theory*. 2000 ISBN 0-7923-6180-6
291. J. Gasser (ed.): *A Boole Anthology*. Recent and Classical Studies in the Logic of George Boole. 2000 ISBN 0-7923-6380-9
292. V.F. Hendricks, S.A. Pedersen and K.F. Jørgensen (eds.): *Proof Theory*. History and Philosophical Significance. 2000 ISBN 0-7923-6544-5
293. W.L. Craig: *The Tensed Theory of Time*. A Critical Examination. 2000 ISBN 0-7923-6634-4
294. W.L. Craig: *The Tenseless Theory of Time*. A Critical Examination. 2000 ISBN 0-7923-6635-2
295. L. Albertazzi (ed.): *The Dawn of Cognitive Science*. Early European Contributors. 2001 ISBN 0-7923-6799-5
296. G. Forrai: *Reference, Truth and Conceptual Schemes*. A Defense of Internal Realism. 2001 ISBN 0-7923-6885-1
297. V.F. Hendricks, S.A. Pedersen and K.F. Jørgensen (eds.): *Probability Theory*. Philosophy, Recent History and Relations to Science. 2001 ISBN 0-7923-6952-1
298. M. Esfeld: *Holism in Philosophy of Mind and Philosophy of Physics*. 2001 ISBN 0-7923-7003-1
299. E.C. Steinhart: *The Logic of Metaphor*. Analogous Parts of Possible Worlds. 2001 ISBN 0-7923-7004-X
300. P. Gärdenfors: *The Dynamics of Thought*. 2005 ISBN 1-4020-3398-2
301. T.A.F. Kuipers: *Structures in Science Heuristic Patterns Based on Cognitive Structures.* An Advanced Textbook in Neo-Classical Philosophy of Science. 2001 ISBN 0-7923-7117-8
302. G. Hon and S.S. Rakover (eds.): *Explanation*. Theoretical Approaches and Applications. 2001 ISBN 1-4020-0017-0
303. G. Holmström-Hintikka, S. Lindström and R. Sliwinski (eds.): *Collected Papers of Stig Kanger with Essays on his Life and Work*. Vol. I. 2001 ISBN 1-4020-0021-9; Pb ISBN 1-4020-0022-7
304. G. Holmström-Hintikka, S. Lindström and R. Sliwinski (eds.): *Collected Papers of Stig Kanger with Essays on his Life and Work*. Vol. II. 2001 ISBN 1-4020-0111-8; Pb ISBN 1-4020-0112-6
305. C.A. Anderson and M. Zelëny (eds.): *Logic, Meaning and Computation*. Essays in Memory of Alonzo Church. 2001 ISBN 1-4020-0141-X
306. P. Schuster, U. Berger and H. Osswald (eds.): *Reuniting the Antipodes – Constructive and Nonstandard Views of the Continuum*. 2001 ISBN 1-4020-0152-5
307. S.D. Zwart: *Refined Verisimilitude*. 2001 ISBN 1-4020-0268-8
308. A.-S. Maurin: *If Tropes*. 2002 ISBN 1-4020-0656-X
309. H. Eilstein (ed.): *A Collection of Polish Works on Philosophical Problems of Time and Spacetime*. 2002 ISBN 1-4020-0670-5
310. Y. Gauthier: *Internal Logic*. Foundations of Mathematics from Kronecker to Hilbert. 2002 ISBN 1-4020-0689-6
311. E. Ruttkamp: *A Model-Theoretic Realist Interpretation of Science*. 2002 ISBN 1-4020-0729-9
312. V. Rantala: *Explanatory Translation*. Beyond the Kuhnian Model of Conceptual Change. 2002 ISBN 1-4020-0827-9
313. L. Decock: *Trading Ontology for Ideology*. 2002 ISBN 1-4020-0865-1

314. O. Ezra: *The Withdrawal of Rights*. Rights from a Different Perspective. 2002
ISBN 1-4020-0886-4
315. P. Gärdenfors, J. Woleński and K. Kijania-Placek: *In the Scope of Logic, Methodology and Philosophy of Science*. Volume One of the 11th International Congress of Logic, Methodology and Philosophy of Science, Cracow, August 1999. 2002
ISBN 1-4020-0929-1; Pb 1-4020-0931-3
316. P. Gärdenfors, J. Woleński and K. Kijania-Placek: *In the Scope of Logic, Methodology and Philosophy of Science*. Volume Two of the 11th International Congress of Logic, Methodology and Philosophy of Science, Cracow, August 1999. 2002
ISBN 1-4020-0930-5; Pb 1-4020-0931-3
317. M.A. Changizi: *The Brain from 25,000 Feet*. High Level Explorations of Brain Complexity, Perception, Induction and Vagueness. 2003 ISBN 1-4020-1176-8
318. D.O. Dahlstrom (ed.): *Husserl's* Logical Investigations. 2003 ISBN 1-4020-1325-6
319. A. Biletzki: *(Over)Interpreting Wittgenstein.* 2003
ISBN Hb 1-4020-1326-4; Pb 1-4020-1327-2
320. A. Rojszczak, J. Cachro and G. Kurczewski (eds.): *Philosophical Dimensions of Logic and Science.* Selected Contributed Papers from the 11th International Congress of Logic, Methodology, and Philosophy of Science, Kraków, 1999. 2003 ISBN 1-4020-1645-X
321. M. Sintonen, P. Ylikoski and K. Miller (eds.): *Realism in Action.* Essays in the Philosophy of the Social Sciences. 2003 ISBN 1-4020-1667-0
322. V.F. Hendricks, K.F. Jørgensen and S.A. Pedersen (eds.): *Knowledge Contributors*. 2003
ISBN Hb 1-4020-1747-2; Pb 1-4020-1748-0
323. J. Hintikka, T. Czarnecki, K. Kijania-Placek, T. Placek and A. Rojszczak † (eds.): *Philosophy and Logic In Search of the Polish Tradition.* Essays in Honour of Jan Woleński on the Occasion of his 60th Birthday. 2003 ISBN 1-4020-1721-9
324. L.M. Vaina, S.A. Beardsley and S.K. Rushton (eds.): *Optic Flow and Beyond.* 2004
ISBN 1-4020-2091-0
325. D. Kolak (ed.): *I Am You.* The Metaphysical Foundations of Global Ethics. 2004
ISBN 1-4020-2999-3
326. V. Stepin: *Theoretical Knowledge*. 2005 ISBN 1-4020-3045-2
327. P. Mancosu, K.F. Jørgensen and S.A. Pedersen (eds.): *Visualization, Explanation and Reasoning Styles in Mathematics*. 2005 ISBN 1-4020-3334-6
328. A. Rojszczak (author) and J. Wolenski (ed.): *From the Act of Judging to the Sentence*. The Problem of Truth Bearers from Bolzano to Tarski. 2005 ISBN 1-4020-3396-6
329. A-V. Pietarinen: *Signs of Logic*. Peircean Themes on the Philosophy of Language, Games, and Communication. 2006 ISBN-10 1-4020-3728-7
330. A. Aliseda: *Abductive Reasoning*. Logical Investigations into Discovery and Explanation. 2006
ISBN-10 1-4020-3906-9

Previous volumes are still available.